GROUNDWATER CONTAMINANT TRANSPORT
IMPACT OF HETEROGENEOUS CHARACTERIZATION

GROUNDWATER CONTAMINANT TRANSPORT

Impact of Heterogeneous Characterization:
A New View on Dispersion

AMRO M.M. ELFEKI
Department of Irrigation and Hydraulics, Faculty of Engineering,
University of Mansoura, Mansoura, Egypt

GERARD J.M. UFFINK
Laboratory of Geotechnics, Faculty of Civil Engineering, Technical University
Delft, Netherlands

FRANS B.J. BARENDS
Laboratory of Geotechnics, Faculty of Civil Engineering, Technical University
Delft, Netherlands, and R & D-division, Delft Geotechnics, Delft, Netherlands

A.A.BALKEMA / ROTTERDAM / BROOKFIELD /1999

Errata

p. 101, Eq. (4.4): K_{xx} and K_{yy} must be interchanged
p. 105, Eq. (4.17): K_{xx} and K_{yy} must be interchanged

Published by
A.A.Balkema, P.O.Box 1675, 3000 BR Rotterdam, Netherlands (Fax: +31.10.4135947)
A.A.Balkema Publishers, Old Post Road, Brookfield, VT 05036-9704, USA (Fax: 802.276.3837)

ISBN 90 5410 665 4

Contents

Contents

Preface

This book is fully based on the dissertation of Dr Amro M.M. Elfeki for his Doctor of Science degree at the Technical University of Delft on the 16th of September 1996[1]. His work was guided and supervised, in close cooperation, by Dr Gerard J.M. Uffink and Prof. Dr Frans B.J. Barends.

The book contains a comprehensive review of existing methods and it offers a new view on the theory of groundwater flow and dispersion of contaminants in natural heterogeneous porous media, using two-dimensional coupled Markov-chain theory and random field theory.

In fact the theory of dispersion, often simplified to Gaussian distribution characteristics, and usually described with one process parameter, the dispersion coefficient, is physically not consistent and therefore, its application in practice is limited. The proof by numerical simulation and physical tests is described in this book.

The consequence is significant. A calculation performed with the standard method using a macro-dispersion coefficient appears to be a folly for naturally heterogeneous porous media. In fact, no realistic answere can be given on the development and transport of local concentrations of contaminants in groundwater. Moreover, the transverse macro-dispersion coefficient is artificial; it has no realistic significance.

Fortunately, it is possible, when using the new method indicated in this book, to outline the area of extent of contamination by groundwater flow in heterogeneous porous media with sufficient accuracy, but in general a precise quantification of the intensity of the concentration inside this area can not be given.

September 30, 1996   *Amro M.M. Elfeki*
Delft, Netherlands   *Gerard J.M. Uffink*
                            Frans B.J. Barends

---

[1]Elfeki, A.M. 1996. Stochastic Characterization of Geological Heterogeneity and Its Impact on Groundwater Contaminant Transport. Dr-Sc Thesis. Geotechnical Laboratory, Faculty of Civil Engineering, TU Delft, Netherlands.

CHAPTER 1

Introduction

1.1 Introduction

Over the last two decades transport of contaminants by groundwater has attracted considerable attention by many researchers from different disciplines working in the field of porous media physics and chemistry. Abundant studies are devoted to this area of research due to the growing concerns about water quality and pollution problems in the biosphere. At present a variety of more or less sophisticated modelling techniques is available for the simulation of subsurface flow and contaminant transport, but reliable predictions can be made only with sufficient knowledge of the geological and hydrogeological parameters of the system. These predictions are of great importance for a more reliable design of groundwater protection schemes, optimal remediation strategies of contaminated aquifers, more adequate preventive measures for vulnerable sites and for decision making. There is, however, often a high uncertainty both on the geometrical structure of the geological units and on the parameter values for the individual units.

Usually, only a limited number of direct measurements of hydrogeological parameters: '*hard data*', such as hydraulic conductivity, porosity, storativity, dispersivity, etc. exists in the form of field observations and laboratory tests, while indirect subjective or qualitative information: '*soft information*', may be more easily available e.g. from geological surveys, intuition, geological interpretations, common experience etc. In general, acquisition of soft data is easier and less expensive. The combination of these types of information is important for improving aquifer characterization and reducing uncertainty.

1.2 Heterogeneity of Natural Formations

Heterogeneity of natural formations can be observed from outcrops and open pits [e.g. Gelhar, 1986; Jussel, et al., 1994]. An example of discrete features of the heterogeneity can be observed in Fig.(1.1a). The figure shows a fluvioglacial deposit consisting of alternating sets of fine silty sand with medium and coarse sand, somewhat contorted. The sedimentary structures are planar and trough crossbedding for the coarse sets and climbing ripples and planar lamination for the finer sets. The sets are overlain by glacial tills (coarse gravel). The hight of the section is about 1.5 m Fig.(1.1b) shows fine to medium sand deposits of a tidal channel at Eastern Scheld, The Netherlands. The dominant sedimentary structures are large sigmoidal sets in beds of up to 1 m in thickness. Individual sets are covered with fine grained material, which merge into a band of finely interlayered sand and fine grained deposits at the toe of the sets. The internal structure comprises inclined lamination and various small scale structures at the toe of the sets.

(a)

(b)

Fig.(1.1) Natural Deposits: (a) Fluvioglacial Deposits in Germany, (b) Tidal Channel Deposits in The Netherlands, by the Courtesy of Kruse.

The continuous nature of the heterogeneity can be observed from measurements of permeability and porosity as shown in Fig.(1.2). The figure presents laboratory measurements of cores from a borehole in Mount Simon sandstone aquifer in Illinois. The transect shows erratic profiles of both permeability and porosity.

From the foregoing pictures and figures the idea arose to develop a more general and systematic methodology to describe the various features of heterogeneity based on geological and parametric information, and their effect on groundwater contaminant transport.

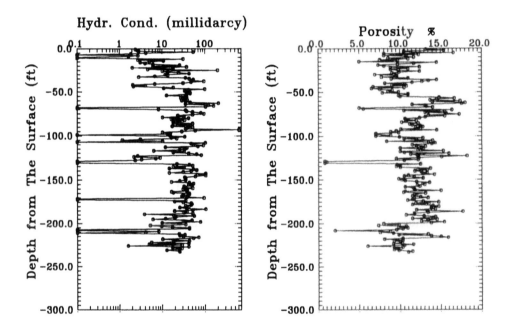

Fig.(1.2) Permeability (millidarcy) and Porosity Space Series from Laboratory Tests, by the Courtesy of Gelhar.

1.3 Objectives of The Research

The objective of this research is to develop a systematic methodology for characterizing geological formations using soft and hard information. Firstly, a stochastic characterization method based on coupled Markov Chain theory is developed to quantify soft information. The soft information can be easily coded by the Markovian methodology. Secondly, a hybrid approach has been developed to merge flow and transport parameters that is derived both from hard data and soft information to enhance the statistical characterization of aquifer heterogeneity.

The hybrid model is a two stages approach: in the first stage, the geological features are modelled stochastically using soft information to describe the geometrical structure of the lithologies with discrete average characteristics. In the second stage, the parameter variability inside each unit is described by correlated continuous statistical variations (stationary random space functions). The usual groundwater flow and transport equations can then be solved in these geological structures. The impacts of the proposed stochastic characterization on the flow and transport are analyzed with single and multi-realizations, and elaborated by numerical and laboratory experiments.

The developed methodologies provide new insight in the various mixing mechanisms that take place in heterogeneous formations, the influence of mega- and macro-scopic variability on flow and transport behaviour, and the impacts of parametric and geological uncertainty on model predictions.

1.4 Scope of The Research

The scope of the study is limited to non-reactive (inert) and single species solutes in steady two-dimensional vertical cross-section groundwater flow systems with a uniform density and miscible contaminant transport. The simulations described in this study use both analytical and numerical models in case of homogeneous medium, whereas, numerical models are used for heterogeneous media. The study is limited to transport by advection and dispersion. For future directions, this methodology could be extended to three-dimensional systems, transient groundwater flows, chemical reactions, multi-species solutes with biological processes, density dependent flow and immiscible contaminants such as LNAPL (Light Non-aqueous Phase Liquid) and/or DNAPL (Dense Non-aqueous Phase Liquid).

1.5 Thesis Layout and Overview of The Contents

The content of this dissertation is as follows:

In Chapter 1 a general introduction addressing objectives, scope and overview of the study is presented.

In Chapter 2 the basic concepts and axioms of a stochastic process and its applications in the field of groundwater mechanics and solute transport are reviewed. The scales of natural variability are given. The common analytical methods to solve stochastic differential equations are described briefly, while the numerical methods are presented with more elaboration. The method of Monte-Carlo sampling is described. A survey of discrete and continuous random field generators has been executed and some commonly used techniques are elaborated. Advantages, disadvantages, and limitations of the techniques used for analytical and numerical stochastic models are evaluated.

In Chapter 3 a new stochastic methodology for characterization of heteroge-neous formation is proposed. The basic principles of classical Markov chain theory are discussed. Then, the theoretical concepts of this methodology are given. Estimation of the model parameters for field applications and its implementation are discussed. Some geological simulations with simple and complex geological settings are displayed. Extension of the methodology to generate spatially distributed hydraulic conductivity fields with prescribed distribution is presented.

Chapter 4 deals with some applications of the developed methodology in the field of groundwater. A full description of the two-dimensional flow model in terms of flow potential and stream function is presented. Some numerical flow simulations are carried out in some hypothetical formations, with characteristics of realistic ones. The rest of the chapter is devoted to the performance of some numerical experiments in analogy to laboratory experiments in order to infer and test statistical properties of the effective hydraulic conductivity of some selected structures. Imperfect stratifications and inclined bedding are considered.

Chapter 5 describes a new hybrid approach. The purpose is to merge soft information and hard data in a stochastic context to enhance the statistical characterization of aquifer heterogeneity and to introduce a compound variability (macroscopic and megascopic variability). Some generated fields are analyzed by the use of the mixture of population model and the variogram method.

Chapter 6 is devoted to single realization numerical experiments of flow and transport in order to study the various mixing mechanisms that take place in heterogeneous formations. In this chapter a numerical transport model is described to simulate single species inert contaminant transport in various geological structures. The transport model reads the Darcy's velocity distribution from the flow model presented in Chapter 4. The transport model is based on a Lagrangian particle tracking algorithm [Uffink, 1990]. The volume-averaged concentration distributions, evolution of plume spatial moments, breakthrough curves and macro-dispersion coefficients are calculated and displayed in a graphical form. The various characterization methodologies (Gaussian fields, Markovian fields and compound fields) are analyzed. Transport in two-scales heterogeneous formations is also investigated.

Chapter 7 presents multi-realizations numerical experiments of solute transport. The purpose of the experiments is to estimate the uncertainty of the transport predictions in terms of ensemble variance over many equally probable realization of the system. These experiments are carried out by Monte-Carlo method. A new approach has been developed to estimate predictions in the presence of geological, parametric and combined uncertainty. The coupled Markov chain model, developed in Chapter 3, allows the quantification of the geological uncertainty for such experiments. New and unexpected results are displayed.

Chapter 8 describes the physical experiments performed on solute transport at laboratory scale. The set-up of the model, parameter estimation of the physical properties of the various glass beads used in the experiments, the method of building artificially heterogeneous aquifers, and the monitoring of the plume behaviour by photographs are described. A deterministic and stochastic numerical simulation of the fluid flow and plume evolution are carried out. The plume spatial moments are also analyzed.

Chapter 9 contains general conclusions and some recommendations for future studies and further development.

CHAPTER 2

Review of The Theory of Stochastic Modelling of Subsurface Porous Flow and Transport

2.1 Introduction

Observations of geological media show significant spatial variability at various scales. Geological formations contain different features of variability such as faults, fractures, lithologies, horizontal layering, inclined bedding, inclusions, lamination and variability of the physical and chemical properties within individual units. Field measurements of hydraulic properties such as hydraulic conductivity, porosity, storativity, dispersivity, etc. exhibit a large degree of spatial variability [Freeze, 1975; Delhomme, 1979; Gelhar, 1986]. The observed variability suggests that it may be useful to describe such parameters in a stochastic context rather than in the traditional deterministic one.

The traditional deterministic approach in which the aquifer properties are represented as a unique local parameter throughout the entire flow domain or represented by multi-layered system each of which can be characterized by distinct parameter is not realistic in many geological settings. In reality, subsurface hydrogeological parameters rarely possess uniform properties; on the contrary, their properties usually vary in a discrete or continuous manner on a multiplicity of scales from one location to another. The often encounters random spatial fluctuations cannot be adequately described by smooth deterministic functions [e.g. Bakr, 1976 and Sudicky, 1986]. On the other hand, aquifer parameters are uncertain. This uncertainty is due to the fact that parameters are measured only at some sampled locations such as selected well locations and depth intervals which are often sparse and/or the intrinsic complexity of the geological process which causes the variability. Groundwater flow and transport are therefore more realistically modelled via the stochastic approach.

The word stochastic has its origin in the Greek adjective στοχαστικος which means skilful at aiming or guessing [Haldorsen, Brand and MacDonald, 1987]. The classical theory of flow through porous media is derived, on deterministic bases, from the principles of continuum mechanics. The principle of mass conservation of flow through a saturated control volume of the porous medium leads to an expression for the continuity through the porous medium in terms of the specific discharge.

By introducing Darcy's law, one can get the governing equation of flow through porous media in terms of pressures. This equation is in the form of an ordinary differential equation in one-dimensional steady flow problems, and in the form of partial differential equations in two- and three-dimensional flow problems.

Stochastic techniques either analytical or numerical are available to solve these differential equations with stochastic parameters. These techniques are the tools to evaluate the effects of spatial variability of the hydrogeological parameters on flow and transport characteristics in porous formations. Before discussion of some selected techniques a brief review of the concepts and properties of stochastic processes is given.

2.2 Basic Concepts, Definitions and Axioms of Stochastic Processes

2.2.1 Definition of Stochastic Processes

A stochastic process may be defined, according to Bartlett [1960, referenced by Krumbein, 1967] 'some possible actual, e.g. physical process in the real world, that has some random or stochastic element involved in its structures'. If a given process operating through time or space is thought of as a system comprising a particular set of states, then in a classical deterministic model the state of the system in time or space can be exactly predicted from knowledge of the functional relation specified by the governing differential equations of the system (deterministic regularity). On the other hand, in the purely stochastic model, the state of the system at any instant or point in time or space is characterized by the underling fixed probabilities of the states in the system (statistical regularity).

A stochastic process can be defined mathematically as a collection of random variables. This definition could be given in a mathematical form as the set $\{[x, Z(x,\zeta_i)], x \in R^n \}$, $i = 1,2,3...,m$ [Marsily, 1986]. $Z(x,\zeta)$ is stochastic process, (random function), x is the coordinates of a point in n-dimensional space, ζ is a state variable (the model parameter), $Z(x,\zeta_i)$ represents one single realization of the stochastic process, $i= 1,2,...,m$ (i: number of realizations of the stochastic process Z), $Z(x_0,\zeta)$ = random variable, i.e., the ensemble of the realizations of the stochastic process Z at x_0, and $Z(x_0,\zeta_i)$= single value of Z at x_0. For simplification the variable ζ is generally omitted and the notation of this stochastic process is $Z(x)$. Fig.(2.1) shows a graphical representation of realizations of a stochastic process.

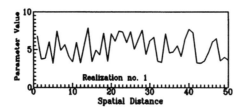

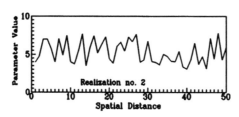

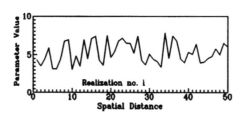

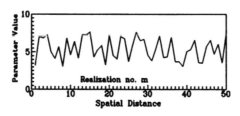

Fig.(2.1) Typical Example of Realizations of Stochastic Process.

2.2.2 Uni-dimensional Stochastic Process

A stochastic process in which the variation of a property of a physical phenomenon is represented in one coordinate dimension is called uni-dimensional stochastic process. The coordinate dimension can be time as in time series, or space as in space series.

2.2.3 Spatial Random Fields

Random fields are multi-dimensional stochastic processes. Some researchers consider random fields as a more general theory from which the uni-dimensional stochastic process is a special case. A random field is defined as a mathematical way to describe spatial variations of properties of a physical phenomenon. These spatial variations can be studied by means of stochastic processes representing these variations in a continuous sense over the space considered or at discrete points in it.

2.2.4 Probabilistic Description of Stochastic Processes

2.2.4.1 Probability Distribution Function (Cumulative Distribution Function)

A single random variable can be defined completely by its probability functions. The probability distribution of a random variable Z is defined by,

$$P(z) = Prob\ \{Z \leq z\} \tag{2.1}$$

where, $Prob\{A\}$ is a probability of occurrence of the event A, and $P(A)$ is a cumulative distribution function of event A, z is a value in deterministic sense. The distribution function is monotonically nondecreasing.

$$\begin{aligned} P(-\infty) &= 0 \\ P(+\infty) &= 1 \end{aligned} \tag{2.2}$$

2.2.4.2 Probability Density Function (PDF)

The density function, $p(z)$, of random variable Z is defined by,

$$p(z) = \lim_{\Delta z \to 0} \frac{Prob\ \{z < Z \leq z + \Delta z\}}{\Delta z} = \frac{dP(z)}{dz} \tag{2.3}$$

Inversely, the distribution function can be expressed in terms of the density function as follows

$$P(z) = \int_{-\infty}^{z} p(z) \, dz \qquad (2.4)$$

$p(z)$ is not a probability, but must be multiplied by a certain region Δz to obtain a probability. $P(z)$ is dimensionless, but $p(z)$ is not. It has dimension of $[z^{-1}]$.

2.2.4.3 Joint Probability Distribution Function

A continuous uni-dimensional stochastic process or random field requires joint functions between all points in space to describe the process completely. It is an extension of single random variable discussed in the previous section. Consider Z as a random vector defined in a vector form as $\{Z_1, Z_2, \ldots Z_n\}^T$, where, $Z_1, Z_2, \ldots$, and Z_n are single random variables. z is described by the joint distribution function of Z as,

$$P(z_1, z_2, \ldots, z_n) = Prob \ (Z_1 \leq z_1, Z_2 \leq z_2, \ldots\ldots, Z_n \leq z_n) \qquad (2.5)$$

This joint distribution function has the same properties as mentioned before, i.e.,

$$\begin{aligned} P(-\infty, \ -\infty, \ -\infty, \ \ldots\ldots, -\infty) &= 0 \\ P(+\infty, \ +\infty, \ +\infty, \ \ldots\ldots, +\infty) &= 1 \end{aligned} \qquad (2.6)$$

2.2.4.4 Joint Probability Density Function

The joint probability density function of Z is defined by,

$$p(z) = \lim_{\substack{\Delta z_1 \to 0 \\ \Delta z_n \to 0}} \frac{Prob \ \{z_1 < Z_1 \leq z_1 + \Delta z_1, \ldots, z_n < Z_n \leq z_n + \Delta z_n\}}{\Delta z_1 \Delta z_2 \ldots \Delta z_n} = \frac{\partial^n P(z)}{\partial z_1 \ldots \partial z_n} \qquad (2.7)$$

Here, z without index is a vector. Inversely, the joint distribution function can be expressed in terms of the joint density as follows,

$$P(z) = \int_{-\infty}^{z} p(z) \, dz = \int_{-\infty}^{z_1} \ldots\ldots\ldots \int_{-\infty}^{z_n} p(z) \, dz_1 \ldots dz_n \qquad (2.8)$$

11

2.2.4.5 Marginal Probability Distribution Function

In general, the marginal probability distribution function of a component Z_l of the random vector Z is obtained from the joint density function by the integration,

$$P(z_1) = Prob(Z_1 < z_1, -\infty < Z_2 < \infty, \dots -\infty < Z_n < \infty)$$

$$= \int_{-\infty}^{z_1} \left[\int_{-\infty}^{\infty} \dots \int_{-\infty}^{\infty} p(z) \ dz_2 \dots dz_n \right] dz_1 \tag{2.9}$$

the term between brackets is the marginal probability density function of the component Z_1, and $P(z_1)$ is called the marginal distribution function of the component Z_1 of the random vector Z.

2.2.4.6 Marginal Probability Density Function

The marginal probability density function is derived from Eq.(2.9) as follows,

$$p(z_i) = \int_{-\infty}^{\infty} \dots \dots \int_{-\infty}^{\infty} p(z) \ dz_1 \ dz_2 \dots dz_{i-1} dz_{i+1} \dots dz_n \tag{2.10}$$

2.2.4.7 Conditional Probability Distribution Function

The conditional probability occurs frequently in the simulation of random fields. In conditional simulations, a field is generated that reproduces observed data at several locations in the field.

The conditional distribution function of one component Z_n of a random vector Z given that the random components at n-1, n-2,........1 have specified values is defined by,

$$P(z_n \mid z_{n-1}, z_{n-2}, \dots, z_1) =$$
$$Prob \ \{Z_n \leq z_n \mid z_{n-1} < Z_{n-1} \leq z_{n-1} + \Delta z_{n-1}, \dots, z_1 < Z_1 \leq z_1 + \Delta z_1\} \tag{2.11}$$

where, the definition $Prob\{A \mid B\}$ is the conditional probability of event A given that, event B has occurred and is defined by,

$$Prob \ \{ A \mid B \} = \frac{Prob \ \{ A \cap B \}}{Prob \{ B \}} \tag{2.12}$$

where, $A \cap B$ is the conjunctive event of A and B.

2.2.4.8 Conditional Probability Density Function

The conditional density function of component Z_n of the random vector Z given that the random components at n-1, n-2, ...1 have specified values is defined by

$$p(z_n \mid z_{n-1}, ..., z_1) = \frac{\partial\, P(z_n | z_{n-1}, ..., z_1)}{\partial z_n} \qquad (2.13)$$

the function $p(z_n \mid z_{n-1}, ...z_1)$ can be expressed in a more convenient form as follows,

$$p(z_n \mid z_{n-1}, z_{n-2}, ...z_1) = \frac{p(z)}{p(z_1, z_2, ..., z_{n-1})} \qquad (2.14)$$

where, $p(z)$, is the joint density function of all the components of the vector Z. It can also be written,

$$p(z) = p(z_1, z_2, z_3, ..., z_n) \qquad (2.15)$$

From Eq.(2.14), Eq.(2.15) one obtains,

$$p(z_n \mid z_{n-1}, z_{n-2}, ...z_1) = \frac{p(z_1, z_2, z_3, ..., z_n)}{p(z_1, z_2, ..., z_{n-1})} \qquad (2.16)$$

If Z_1, Z_2, ..., and Z_n are independent random variables then the joint density function is the multiplication of the marginal density function of the individual random components. This can be expressed as follows,

$$p(z_1, z_2, z_3, ..., z_n) = p(z_1) \cdot p(z_2)....... p(z_n) \qquad (2.17)$$

So, in conclusion, for independent random variables the following holds,

$$p(z_n \mid z_{n-1}, ...z_1) - p(z_n).$$

2.2.5 Statistical Properties of Stochastic Processes

2.2.5.1 Spatial or Temporal Statistical Properties

The statistical properties of a stochastic process on one single realization over specified interval in space (in case of space series, i.e., the process is variable in space) or in time (in case of time series, i.e., the process is variable in time) are called spatial or temporal statistical properties. In this section, attention to the spatial processes will be given, because in geological situations, the observation points are ordered in space not in time. The mathematical expressions of spatial properties of the spatial processes are given below.

(i) Spatial Average

In a continuous space:

$$\overline{Z}_i = \frac{1}{|v|} \int_{v(x')} Z_i(x) \, dx \tag{2.18}$$

where, $v(x')$ is the specified length, area or volume (for one, two or three dimensional space respectively) centred at x' of measure $|v|$ and index i is the i-th realization.

In a discrete space:

$$\overline{Z}_i \simeq \frac{1}{n} \sum_{j=1}^{n} Z_i(x_j) \tag{2.19}$$

where, n is the number of discrete points discretizing the volume v, index j is the j-th point on volume v.

(ii) Spatial Mean Square

In a continuous space:

$$\overline{Z_i^2} = \frac{1}{|v|} \int_{v(x')} Z_i^2(x) \, dx \tag{2.20}$$

In a discrete space:

$$\overline{Z_i^2} \simeq \frac{1}{n} \sum_{j=1}^{n} Z_i^2(x_j) \tag{2.21}$$

(iii) Spatial Variance

The variance is a measure of the dispersion of the process about its average value; in a sense, it gives the average fluctuations of the process about a statistically mean value.

In a continuous space:

14

$$Var[Z_i] = \sigma^2_{Z_i} = \overline{\left[Z_i(x) - \overline{Z_i}\right]^2} = \frac{1}{|v|} \int_{v(x')} \left[Z_i(x) - \overline{Z_i}\right]^2 dx \quad (2.22)$$

In a discrete space:

$$\sigma^2_{Z_i} \approx \frac{1}{n-1} \sum_{j=1}^{n} \left[Z_i(x_j) - \overline{Z_i}\right]^2 \quad (2.23)$$

or after some elaboration:

$$\sigma^2_{Z_i} \approx \overline{Z_i^2} - (\overline{Z_i})^2 \quad (2.24)$$

If the number n is large, somtimes the denominator $n-1$ in Eq.(2.23) is simply replaced by n, (see Eq.(7.5) and Eq.(7.6)). In that case Eq.(2.24) is exact.

(iv) Spatial Covariance

The covariance is a measure of the mutual variability of a pair of realizations; or in other words, it is the joint variation of two variables about their common mean. It can be expressed mathematically as follows.
In a continuous space:

$$Cov(Z_i(x+s), Z_i(x)) =$$
$$\frac{1}{|v|} \int_{v(x')} \left[Z_i(x+s) - \overline{Z_i(x+s)}\right] \left[Z_i(x) - \overline{Z_i(x)}\right] dx \quad (2.25)$$

where, s is called the spatial lag.
In a discrete space:

$$Cov(Z_i(x+s), Z_i(x)) \approx$$
$$\frac{1}{n(s)} \sum_{j=1}^{n(s)} \left[Z_i(x_j+s) - \overline{Z_i(x_j+s)}\right] \left[Z_i(x_j) - \overline{Z_i(x_j)}\right] \quad (2.26)$$

where, $n(s)$ is the number of points with lag s.

2.2.5.2 Ensemble Statistical Properties (Mathematical Expectation)

The average of statistical properties over all possible realizations of the process at a given point on the process axis are called the ensemble statistical properties.

The mathematical expressions of ensemble properties are

(i) Ensemble Average

$$E \{Z(x_o)\} = \int_{-\infty}^{\infty} z(x_o) \, p(z) \, dz \qquad (2.27)$$

where, x_o is the coordinate of a given point on the space axis, $p(z)$ is probability density function of the process $Z(x)$ at location x_o, and $E\{.\}$ is the expected value operator.

For finite number of realizations:

$$E \{Z(x_o)\} \approx \frac{1}{m} \sum_{i=1}^{m} Z_i(x_o) \qquad (2.28)$$

where, m is the number of realizations.

(ii) Ensemble Mean Square

The mathematical definition:

$$E \{Z^2(x_o)\} = \int_{-\infty}^{\infty} z^2(x_o) \, p(z) \, dz \qquad (2.29)$$

For finite number of realizations:

$$E \{Z^2(x_o)\} \approx \frac{1}{m} \sum_{i=1}^{m} Z_i^2(x_o) \qquad (2.30)$$

(iii) Ensemble Variance

$$\sigma^2_{Z(x_o)} = E \{[Z(x_o) - E\{ Z(x_o)\}]^2\} = \int_{-\infty}^{\infty} (Z(x_o) - E \{Z(x_o)\})^2 \, p(z) \, dz \quad (2.31)$$

For finite number of realizations:

$$\sigma^2_{Z(x_o)} \approx \frac{1}{m-1} \sum_{i=1}^{m} [Z_i(x_o) - E \{Z(x_o)\}]^2 \qquad (2.32)$$

or,

16

$$\sigma^2_{Z(x_o)} = E \{Z^2(x_o)\} - \left(E \{Z(x_o)\}\right)^2 \qquad (2.33)$$

(iv) Ensemble Covariance

$$Cov(Z(x+s), Z(x)) = \int\int_{-\infty}^{\infty} p(z(x+s), z(x))dz(x+s)dz(x) \qquad (2.34)$$

where, $p(z(x+s),z(x))$ is the joint probability density function of the process $Z(x)$ at locations $x+s$ and x.

For finite number of realizations:

$$Cov(Z(x+s), Z(x)) \approx$$
$$\frac{1}{m} \sum_{j=1}^{m} \left[Z_j(x+s) - E \{Z_j(x+s)\}\right].\left[Z_j(x) - E \{Z_j(x)\}\right] \qquad (2.35)$$

2.2.5.3 Stationarity, Non-stationarity, Intrinsic Hypothesis and Ergodicity

(i) Stationarity (Statistical Homogeneity)

Stationarity is a statistical property describing the state of variability of the stochastic process. The stationarity can be tested for all moments of the stochastic process. The stochastic process is said to be second-order stationary if the mean value of a stochastic process is constant at all points in the field, that means, the mean does not depend on the position. This can be expressed mathematically as,

$$E\{Z(x)\} = \mu_Z \qquad (2.36)$$

and if the covariance of a stochastic process depends only on the difference between the position vectors of two points $(x_i\text{-}x_j)= s_{ij}$ the separation vector, and does not depend on the position vectors x_i and x_j themselves. This can be expressed mathematically as,

$$Cov(Z(x_i), Z(x_j))=E\{[Z(x_i)-E\{Z(x_i)\}]\ [Z(x_j)-E\{Z(x_j)\}]\}=Cov(s) \quad (2.37)$$

this implies that the variance is independent of x,

$$Var[Z(x)] = Cov(0) = \sigma^2_Z \qquad (2.38)$$

In the physical sense, one thinks of a medium whose variability is the same throughout the formation of interest, so that the covariance is independent of the position but depends on the separation vector [Bakr et al., 1978].

(ii) Non-stationarity

A stochastic process is called non-stationary, if the moments of the process are variant in space, i.e., from one position to another. On other words, the moments not only depends on the separation vector, but also, on the position of the point in space.

(iii) Intrinsic Hypothesis

This is another statistical property which is weaker than the second order stationarity. The intrinsic hypothesis assumes that even if the variance of $Z(x)$ is not finite, the variance of the first-order increments of $Z(x)$ is finite and these increments are themselves second-order stationary. This hypothesis postulates that: (1) the mean is the same everywhere in the field; and (2) for all distances, s, the variance of the increments, $\{Z(x+s)-Z(x)\}$ is a unique function of s so independent of x. A stochastic process that satisfies the stationarity of order two also satisfies the intrinsic hypothesis, but the converse is not true. The above two properties can be expressed by writing,

$$E \{Z(x+s)-Z(x)\} = 0 \tag{2.39}$$

$$Var \{Z(x+s)-Z(x)\} = 2\ \gamma(s) \tag{2.40}$$

where, $\gamma(s)$ is called the semi-variogram,
or Eq.(2.40) may be written as,

$$\gamma(s) = \frac{1}{2}E \{[Z(x+s)-Z(x)]^2\} \tag{2.41}$$

which means, that $2\gamma(s)$ is the mean squared difference for two points separated by a distance s.

From practical point of view, the intrinsic hypothesis is appealing, because it allows the determination of the statistical structure, without demanding the prior estimation of the mean. Furthermore, for a stationary random process, where both a covariance and a sime-variogram are exist, it is easy to show the relationship between them as,

$$\gamma(s) = Cov(0) - Cov(s) \qquad (2.42)$$

(iv) Ergodicity

Ergodicity is a statistical property which implies that the statistics of a single realization in space (spatial statistics) are equivalent to the ensemble of all possible realizations (ensemble statistics). In other words, by observing the variability in space of a property from one realization in enough detail, it is possible to determine the probability distribution function of the random process for all possible realizations. This equivalence is achieved when the size of the space domain is sufficiently large or tends to infinity.

2.2.5.4 Statistical Isotropy and Anisotropy

A multi-dimensional stochastic process is said to be isotropic, if the process does not have a preferred direction, i.e., the variability in the process is the same in all directions. On the other hand, the process is said to be anisotropic, if the variability changes from one direction to another.

2.2.6 Real (Lag) Domain Representation of Stochastic Processes

Properties of stationary stochastic processes may be represented in a lag domain either as an auto-correlation function of the lag s, or as cross-correlation function of s. The diagram used to display this function is called correlogram which represent the auto-correlation coefficients versus the lag s between the auto-correlated values of the process.

2.2.6.1 Spatial Auto-Correlation

It is a measure of the spatial correlation structure of a process. It gives the degree to which a process is correlated with itself as a function of separation lag. The auto-correlation of stochastic process $Z(x)$ is expressed mathematically by,

$$\rho_{ZZ}(s) = \frac{Cov(Z(x+s), Z(x))}{\sigma^2_z} \qquad (2.43)$$

The auto-correlation function has the following properties.

First property:

$$\rho_{ZZ}(0) = 1 \qquad (2.44)$$

that means, the correlation coefficient between a point and itself is 1 i.e. perfect correlation.

Second property:

$$\rho_{ZZ}(\infty) = 0 \qquad (2.45)$$

that means, the correlation coefficient between two points far enough is zero i.e. the process is uncorrelated when the lag becomes very large and it is said to have no memory for what occurred prior to that lag.

Third property:

$$\rho_{ZZ}(s) = \rho_{ZZ}(-s) \qquad (2.46)$$

that means, the auto-correlation function is an even function (symmetric about the vertical axis).

2.2.6.2 Spatial Cross-Correlation

The spatial cross-correlation represents a relation between two stochastic processes. It defines the degree of which two stochastic process are correlated as a function of separation lag. The cross-correlation is defined mathematically as the expected cross-correlation coefficient for lag s, either positive or negative or zero, for two processes $Z(x)$ and $Y(x)$. This expressed mathematically as,

$$\rho_{ZY}(s) = \frac{Cov(Z(x+s),\ Y(x))}{(\sigma^2_Z\ \sigma^2_Y)^{1/2}} \qquad (2.47)$$

2.2.6.3 Integral Scale and Correlation Scale

(i) Integral Scale

The integral scale I_z of autocorrelation function is defined by Dagan [1982] in analogy to Lumley and Panofsky [1964] as,

20

$$I_z = \int_0^\infty \rho_{ZZ}(s) \, ds \qquad (2.48)$$

which implies that the average distance over which the process is autocorrelated in space. For practical applications, the integration is calculated over a certain limits $[0, S_o]$ where, S_o is the smallest value of s at which the autocorrelation function becomes practically zero.

(ii) Correlation Scale (Range)

The correlation scale is defined as the distance over which the process is autocorrelated in space. It is calculated as the distance at which the autocorrelation function tends to zero. There are various ways [e.g. Smith, et al., 1979 a,b], some authors suggest the threshold value taken as e^{-1} to others [e.g. Gelhar, 1976]. In case of 2D isotropic exponential autocorrelation function is defined by,

$$\rho_{ZZ}(s) = e^{-\frac{|s|}{\lambda}} \qquad (2.49)$$

Adopting the threshold value e^{-1}, one finds that the integral scale is related to the correlation length by the formula,

$$I_z = \frac{\lambda}{\sqrt{2}} \qquad (2.50)$$

2.2.7 Spectral (Frequency) Domain Representation of Stochastic Processes

Properties of stochastic processes can also be represented in the frequency domain, relating the square of amplitude of each sine or cosine component fitting the process versus ordinary frequency or its angular frequency or wave number. In this respect, the stochastic process is considered as made up of oscillations of all possible frequencies. The diagram used for this presentation is called priodgram.

2.2.7.1 Auto-Power (Variance) Spectral Density Function (*PSD*)

The term power is commonly seen in the literature. Its origin comes from the field of electrical and communication engineering, where power dissipated in an electrical circuit is proportional to the mean square voltage applied, hence the term power is used. The adjective spectral denotes a function of frequency. The concept of density comes from the division of the power (variance) of an infinitesimal frequency interval by the width of that interval. The power spectrum describes the distribution of power (variance) with frequency of the random processes, and as such is real and non-negative. The auto-power spectrum (spectral density function) for a process $Z(x)$ is given by,

$$S_{ZZ}(\omega) = \lim_{L \to \infty} \frac{1}{L} \overline{[z(\omega).z^*(\omega)]} = \lim_{L \to \infty} \frac{1}{L} \overline{|z(\omega)|^2} \qquad (2.51)$$

where, $z(\omega)$ is Fourier transform of the process $Z(x)$, which is expressed as,

$$z(\omega) = \frac{1}{2\pi} \int_{-\infty}^{\infty} Z(x)e^{-i\omega x} \, dx \qquad (2.52)$$

and $z^*(\omega)$ is the conjugate of $z(\omega)$ and ω is the angular frequency vector.

2.2.7.2 Cross-Power (Variance) Spectral Density Function (*Cross-PSD*)

The *cross-PSD* is defined between a pair of stochastic process. *Cross-PSD* is in general complex. The magnitude of the *cross-PSD* describes whether frequency components in a process are associated with large or small amplitudes at the same frequency in another process, and the phase of the *cross-PSD* indicates the phase lag or lead of one process with respect to the other one for a given frequency component. This expressed mathematically as,

$$S_{ZY}(\omega) = \lim_{L \to \infty} \frac{1}{L} \overline{[z(\omega).y^*(\omega)]} \qquad (2.53)$$

where, $y^*(\omega)$ is the conjugate of $y(\omega)$, and $y(\omega)$ is Fourier transform of the process $Y(x)$ which can be expressed by Eq.(2.52).

2.2.7.3 Relation between Covariance Functions and Spectral Density Functions

The covariance functions and spectral density functions are Fourier transform pairs. This can be expressed in mathematical forms using Wiener-Khinchin relationships,

$$S_{ZZ}(\omega) = \frac{1}{2\pi} \int_{-\infty}^{\infty} C_{ZZ}(s)e^{-i\omega s} \, ds \qquad (2.54)$$

and its inverse is,

$$C_{ZZ}(s) = \int_{-\infty}^{\infty} S_{ZZ}(\omega)e^{i\omega s} \, d\omega \qquad (2.55)$$

The variance of the stochastic process can be calculated from the spectrum using the following relation,

$$C_{ZZ}(0) = \int_{-\infty}^{\infty} S_{ZZ}(\omega) \, d\omega = \sigma_Z^2 \qquad (2.56)$$

For *cross-PSD* and cross-correlation these relations are,

$$S_{ZY}(\omega) = \frac{1}{2\pi} \int_{-\infty}^{\infty} C_{ZY}(s)e^{-i\omega s} \, ds \qquad (2.57)$$

and its inverse is,

$$C_{ZY}(s) = \int_{-\infty}^{\infty} S_{ZY}(\omega)e^{i\omega s} \, d\omega \qquad (2.58)$$

Eq.(2.51) to Eq.(2.58) are used to transform the analysis of the stochastic process form space domain to frequency domain and vice versa. These relations are used frequently in the analytical study of stochastic processes.

2.3 Scales of Natural Variability (Heterogeneity)

Heterogeneity can be found at various scales. Definitions of such scales differ from author to author. According to [Weber, 1986] these scales are described as:

(i) Microscopic Scale: this scale of variability is over the grains and pores. It is in order of mm. At this scale flow inside the pores and between the grains is governed by the Navier-Stokes equations.

(ii) Macroscopic Scale: at this scale one is interested in cores and samples which contains many grains and pores. This variability is the average of the various microscopic variabilities. Darcy's law is emerged at this scale from the average over the enormous complexity of the Navier-Stockes equations governing the flow at the pore scale. This variability is known also by Dagan [1986] as Darcy or laboratory scale where one is interested in material properties, like, porosity, permeability, and dispersivity. This scale is in order of m which is also named by Bear [1979] the representative elementary volume (*REV*).

(iii) Megascopic Scale: this scale describes the internal architecture of reservoir

units and lithologies. In this scale one is interested in the dimensions, shapes, orientations and spatial disposition of lithofacies. This kind of variability is sometimes called field or local scale variability which is in order of 100 m.

(iv) Gigascopic Scale: at this scale one is interested in regional geological features such as faults, fractures, aquifer size, depositional environments or regional tectonic events. This scale is also known as regional or formation scale which is in order of km.

Several subdivision within these scaling classes are possible that causes the difference between one author to another. The difference between the type of reservoir like hydrocarbon reservoirs and groundwater basins, where some specific features can appear in one reservoir, influence these definitions as well. Fig.(2.2) shows the scales of variability according to Weber [1986].

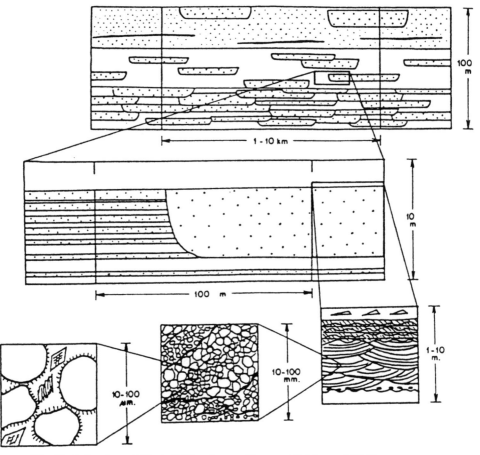

Fig.(2.2) Scales of Variability According to [Weber, 1986].

2.4 Site Characterization and Probabilistic Description

The complex structure of the geological formations has attracted considerable attention from geologists, geostatisticians, hydrogeologists, oil reservoir engineers, civil engineers, mining engineers, environmental engineers and others who are interested in geoscience. The literature on techniques to describe heterogeneity is abundant and rapidly increasing.

2.4.1 Deterministic Approach

The deterministic approach, which can be described simply as drawing the most probable picture of the formation units based on some kind of interpolations between well logs utilizing geological experience and technical background from practitioners. The produced picture is considered as a deterministic one. The final user of that picture may rely on it as a subjectively certain picture of the subsurface.

2.4.2 Stochastic Approach

The basic idea behind the stochastic approach in site characterization is that a large number of synthetic geological structures are generated based on a stochastic description of the system. In case of single stochastic simulation, the geological judgement should be involved to decide for the most probable image of the geological system based on geological experience, intuition, common sense etc.. However, in case of a multi-realization approach, Monte-Carlo approach (see section 2.5.2) is applied to estimate the uncertainty in the output (response) variables. A comprehensive review of the stochastic nature of reservoirs can be found in the literature of oil reservoirs, such as Haldorsen, Brand and MacDonald [1987], and Fayers and Hewett [1992].

In the stochastic approach two lines of thought are found in the literature. The first is the facies or discrete models and the second is the continuous models. This kind of classification can also be viewed as description of the different scales mentioned in section 2.3. For instance, the discrete model is aiming to characterize megascopic scales and the continuous models are used to characterize the macroscopic sales of variability. Description of these models is given next.

2.4.2.1 Mosaic Facies (Discrete) Models

These models focus on the geological description of the discrete features of natural formations. In this approach one is aiming to construct formation geological units, its geometric characteristics, lithologies, units dimensions (lengths, thicknesses, and widths), orientations and frequency of occurrence, etc.. There is a variety of techniques for simulating geological patterns. These models are often used by petroleum geologists. One may distinguish two types:

(i) Object-based models, Boolean models or Binary models:
In these models, one may consider two states or phases e.g. sand and shale formation. Sand bodies are generated in the foreground while the background is shale or vice versa. The shape of the objects can be rectangles, ellipses or any other shape. The objects are distributed randomly in space with random sizes. There are two main parameters needed for such models, the density of the random objects per unit of volume and the statistical distribution of the sizes of the objects. The orientation of the objects with respect to the horizontal can also be considered in these models. Boolean models are very simple and flexible. The limitations come from the randomness in the location of the objects in space and the ignorance of the spatial connectivity between the objects in space. More detailed information can be found in a recent work by Chessa [1995]. The limitations of the object-based models motivated the development of the following models.

(ii) Sequential-based models:
These types of models can be divided broadly into two categories: (*a*) the first category is based on the theory of two-point auto and cross-covariance description of the spatial process discussed in section 2.2.6.1; (*b*) the second category is based on a local conditional probability description of the spatial process (see section 2.2.4.7).

A model based on the first category is the '*Sequential Indicator Simulation*' proposed by Journel [1989]. A reproduction of all facies proportions and auto and cross covariances could be ensured by this method. From a practical point of view, the only difficulty of this method is the inference of reliable covariance functions from the available data. Models based on the second category are:

'*Classical Markov Chain Model*'. In this approach the stratigraphic sequence of geological units is described in terms of transition probability matrix. A one-dimensional Markov chain has been used by Krumbein [1967]. The sequence of layers is random but conditional on the preceding layer. A generalization of this approach is possible where the conditional dependence includes layers earlier than the preceding layer. For more details about this approach see Harbaugh and Bonham-Carter [1970]. A complete description of the Markov chain theory is presented in the next chapter (section 3.3). The limitation of this method is that it only characterize the variability in sedimentological sequence in vertical direction. An extension of this method into higher dimensions is the core of the next chapter where a new methodology is proposed.

'*Poisson Random Lines Model*', its theoretical work has been provided by Switzer [1965, referenced by Lin and Harbaugh, 1984] and has been applied by Lippman [1973]. The lines are used to represent boundaries between different soils. In this approach an estimation of a parameter governing the Poisson distribution is required. The parameter is used to determine the number of random lines used in the simulation step. Attempts have been done to estimate this parameter from the transition probabilities, but the question of the most appropriate procedure for estimating this parameter has not been resolved. Some attempts have been done to use this model, but no satisfactory results are found comparable with natural geological formations.

'Markovian Random Field Model' used by Cross and Jain [1983] to model 2D surface texture of natural materials. This method was originally developed for the application in the field of image processing. There is a similarity between image description and a reservoir description. It seems that such an approach is well suited to the characterization of underground reservoirs. The method does not use variogram or auto-correlation to describe the relationships between neighbouring locations, but it is based on the theory of conditional probability. The idea in this method is that the probability of a given state being present at a given node on a lattice depends only on the states in the immediate neighbourhood of the simulated grid point. Realizations of Markovian fields can be generated iteratively using a simulated annealing algorithm or Metropolis algorithm. Cross and Jain [1983] applied Metropolis algorithm to simulate Markovian fields. The Metropolis algorithm works briefly as follows. The states in the systems are generated by arbitrary distribution over the lattice. Two grid points are selected at random from the lattice. Then, simple exchange of grid points states based on conditional probabilities is performed by Metropolis algorithm. After each trial step in Metropolis algorithm a new permutation of the states of the grid points is created. The procedure continues iteratively until the marginal and the transitional probabilities of the states in the system are stabilized. The method seems promising in the field of reservoir characterization but the disadvantages of this method are its computer-intensive iterative procedure required to achieve equilibrium of the system and the difficulty to perform conditioning to honour measurements at their location. In the next chapter a new methodology has been developed based on coupled Markov chain theory which overcomes the disadvantages mentioned in this method to some extent.

2.4.2.2 Continuous Models

The continuous models are another way of describing heterogeneity. These models focus on rock property or parametric variability to describe the local variations of certain parameters (hydraulic conductivity, porosity, dispersivity... etc.). These types of models are frequently used in the field of stochastic subsurface hydrology. There are several methods for the generation of stochastic fields. Some of these methods are: *'Multi-variate Method'* developed by Iman [1980, referenced by Peck, et al., 1988], *'Nearest Neighbour Method'* developed by Whittle [1954], *'Turning Bands Method'* developed by Matheron [1973], *'Spectral Methods'* by Shinozuka and Jan [1972] and Mejia and Rodriguez-Iturbe [1974], *'Fast Fourier Transform Method'* by Borgman, [1984]; and Gutjahr [1989] and *'Source Point*

Method' by Ghori, Heller and Singh [1992]. The first three methods are discussed in detail in this chapter (see section 2.5.3.2). The other methods are based more or less on the same background. Descriptions of these methods, their advantages, disadvantages limitations and algorithms for implementation are presented in the following sections [see also Elfeki, 1993]. All these methods are based on the theory of regionalized variables developed by Matheron [1971] and they are used to generate realizations of stationary random fields of the model parameters.

2.5 Stochastic Differential Equations (*SDEs*)

Differential equations for random functions (stochastic processes) arise in the investigation of numerous physical and engineering problems. When certain functions, coefficients, parameters and boundary or initial values in classical differential equation (*DE*) are random, the differential equation is called stochastic differential equation (*SDE*). Some selected techniques for solving *SDEs* of flow and transport are compiled in Fig.(2.3). The following discusses these methods.

2.5.1 Analytical Approaches

There are some analytical approaches for solving stochastic differential equations. The main advantage of these approaches is that one can get insight into the effect of parametric variability on a certain dependent variable in the governing equation. One may obtain closed form solutions of the problem or semi-analytical solutions with some numerical quadrature. The usefulness of these analytical methods is limited to relatively simple cases. On the other hand, a large number of assumptions is necessary made in order to find the solution. The results of these analytical methods are the first two moments of the output stochastic variables given the first two moments of the input stochastic parameters. The input stochastic parameters must be statistically homogeneous (second order stationarity see section 2.2.5.3).

2.5.1.1 Perturbation Method

One of the analytical methods for solving stochastic differential equations is the perturbation method. This method has long been recognized in applied mathematics in general and in groundwater field in particular. The principle of this method is simple. The parameter,Y, (e.g. conductivity) and the variable, Φ, (e.g. head) can be expressed in a power series expansion as,

$$\Phi = \Phi_o + \beta\Phi_1 + \beta^2\Phi_2 \ldots\ldots \tag{2.59}$$

$$Y = Y_o + \beta Y_1 + \beta^2 Y_2 \ldots \ldots \tag{2.60}$$

where, β is a small parameter (smaller than unity). These expressions are introduced in the differential equations of the system to get a set of equations in terms of zero- and higher-order expressions of the factor β. The equation that is in terms of zero β corresponds to the mean head. The equation that is in terms of first-order of β corresponds to the head perturbation. In practice, only two or three terms of the series are usually evaluated. This method has been used for solving stochastic differential equations governing groundwater flow by Gutjahr et al. [1978] and Ababou, et al. [1990] and for solving transport equations by Tang and Pinder [1977].

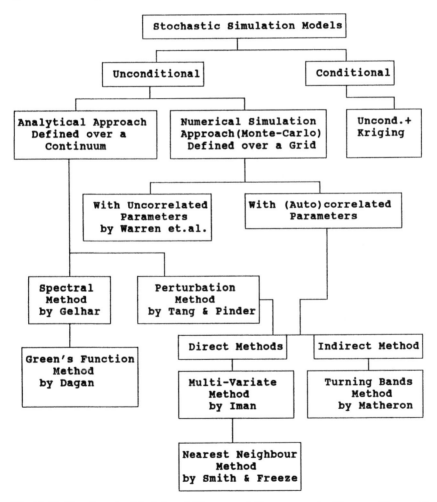

Fig.(2.3) Stochastic Models of Flow and Transport in Porous Formations.

2.5.1.2 Spectral Method

This method is based on studying the spatial variation of the stochastic parameter in the continuum sense as a random field characterized by theoretical spatial covariance functions in real (space) domain or by spectral density functions in the frequency domain.

This method has been applied to the groundwater flow problems by Gelhar and his coworkers [Gelhar, 1977], [Gutjahr and Gelhar, 1981] and [Gelhar, 1986]. A brief description of the method is as follows. The dependent variable and parameter in a stochastic differential equation are represented in terms of its mean or expected value denoted with an angle brackets, and some fluctuations around the mean denoted by a prime, as follows:

$$\Phi = \langle \Phi \rangle + \dot{\Phi} \tag{2.61}$$

$$Y = \langle Y \rangle + \dot{Y} \tag{2.62}$$

where, Y is also written as the perturbed parameter, $\langle Y \rangle$ is the mean or expected value of the parameter, $E\{Y\}$, $\dot{Y}$ is a perturbation around the mean value of the parameter, so $E\{\dot{Y}\} = 0$. Similarly, Φ is the perturbed variable, $\langle \Phi \rangle$ is the mean or expected value of the variable, $E\{\Phi\}$, and $\Phi`$ is a perturbation around the mean value of the variable, $E\{\Phi`\} = 0$.

Two assumptions are made. The first is that the perturbations are relatively small compared to the mean value, so that second order terms involving products of small perturbations can be neglected. The second is that the stochastic inputs parameters and the outputs variables are second order stationary so that they can be expressed in terms of a representation theorem. Introducing Eq.(2.61) and Eq.(2.62) into the differential equation governing the phenomenon under study (e.g. flow or transport) and taking the expected value of the equation, this results in two new equations, one for the first moment (mean) and the other for the perturbations. The first is an ordinary deterministic differential equation, which can be solved analytically to get the solution for the mean value of the dependent variable as a function of the mean value of the independent variable (parameter). The second equation is transformed in the spectral domain by using Fourier-Stieltjes [Lumley and Panofsky, 1964]. The following integral transformation is used,

$$\dot{Y} = \int_{-\infty}^{\infty} e^{ikx} \, dZ_Y(k) \tag{2.63}$$

$$\dot{\Phi} = \int_{-\infty}^{\infty} e^{ikx} \, dZ_{\Phi}(k) \qquad (2.64)$$

where, k is wave number vector, x is space dimension vector, $Z(k)$ is a random function with orthogonal increments, i.e., nonoverlapping differences are uncorrelated and $dZ(k)$ is complex amplitudes of the Fourier modes of wave number k. The spectral density function $S_{YY}(k)$ of $\dot{Y}$ is related to the generalized Fourier amplitude, dZ_Y by

$$E\{dZ_Y(k_1).dZ_Y^*(k_2)\} = 0, \qquad if \quad k_1 \neq k_2 \qquad (2.65)$$

$$E\{dZ_Y(k_1).dZ_Y^*(k_2)\} = S_{YY}(k_1)dk, \qquad if \quad k_1 = k_2 \qquad (2.66)$$

The asterisk, *, denotes the complex conjugate. By using the above representation and substituting them into the stochastic differential equation of the perturbation, one can get the spectrum of the variable as a function of the spectrum of the parameter.

The spectral density function is the Fourier transform of its auto-covariance function, which can be expressed mathematically as follows:

$$S_{\Phi\Phi} = \frac{1}{2\pi} \int_{-\infty}^{\infty} e^{-iks} \, C_{\Phi\Phi}(s) \, ds \qquad (2.67)$$

where, s = vector lag of the auto-covariance function.

This relationship that the power spectral density is the Fourier transform of the auto-correlation function is sometimes referred to as the Wiener-Khinchin theorem [Kay, 1988]. By using inverse Fourier transform, one can also write

$$C_{\Phi\Phi}(s) = \int_{-\infty}^{\infty} e^{-iks} \, S_{\Phi\Phi}(k) \, dk \qquad (2.68)$$

Then, with the above integral it is possible to calculate analytically the auto covariance function of the variables and thus its variance i.e.

$$\sigma_{\Phi}^2 = Cov(0) \qquad (2.69)$$

The spectral approach is the focus of the recent book by Gelhar [1993].

2.5.1.3 Green's Function Approach

An alternative solution method, the Green's function approach, has also been used to analyze the stochastic partial differential equations of flow through porous media. This method was developed by Chirlin and Dagan [1980]. The basic concept of this method is constructing a hypothetical porous medium from a set of spherical blocks in case of three-dimensional problems and cylindrical blocks in case of two-dimensional problems with different conductivities and sizes. This model is sometimes called embedded matrix method or selfconsistent approximation. In this approach, the Green's function is easy to obtain and it allows to express the variable (head) as linear integral transform of the para-meter (conductivity). The flow equations for both steady and unsteady flow and for both finite and infinite domains can be solved analytically with this approach.

2.5.1.4 Some Other Recent Methods

The perturbation approaches are useful tools but its accuracy for large variances is questionable. Therefore limitations of the previous approaches have motivated the development of alternative methods for solving *SPDEs*. Some of the latest methods are:

(1) '*Integral Equations Approach*' where, the differential equations are transformed into integral equations [Serrano, 1988]. One of the advantages of this approach is that the boundary conditions are introduced in the integral equation right from the beginning instead of being introduced at the end, as used to be, in the solution of *PDEs*. This approach is similar to the one used by Dagan in Green's function approach. It is not yet clear if the solution of these integral equations would be easer than the original *PDEs*.

(2) '*Neumann Series Expansion Approach*'. This approach is similar to the one used with power series expansion discussed earlier (section 2.5.1.1). Zeitoun and Braester [1991] used this approach to study the stochastic flow equation. They claimed that the method can handle larger variabilities in conductivity than the classical methods.

(3) '*The Semigroup Approach*' developed by Serrano and Unny [1987] seems promising for large contrasts in conductivity but this has not yet been established.

(4) '*Diagrammatic or Graphic Perturbation Approaches*' [Christakos et al., 1993]. This is a completely different approach from the presented methods. It is claimed that graphic visualizations of the underlying flow processes allow previously undetected features to be seen and can yield more general and accurate results than previous methods.

2.5.2 Numerical Simulation Approaches (Monte-Carlo Simulation)

Monte-Carlo method is first proposed by Von Neumann and Ulam. The name 'Monte-Carlo' arises from the random character of the method and the famous casino in Monaco province, France. Monte-Carlo strategy is appropriate for a broad class of problems in statistical and quantum mechanics. More general detailed presentations of the method can be found in Hammersly and Handscomb [1964]. The idea of the simulation is to build a numerical model of the aquifer that has the same statistical characteristics as the real one and to simulate on the model the same phenomena (porous flow, transport,..etc.) that can occur in reality.

Monte-Carlo method is a most powerful tool in simulating stochastic phenomena, while few assumptions are required. It is the easiest method to understand. One can easily handle uncertainty in model parameters when the number of parameters is more than one, which is otherwise very difficult to handle by analytical methods. One can study systems which not yet tractable with analytical methods. Monte-Carlo techniques can handle easily large input variances which practically cannot be done by analytical methods. Also, the problems of bounded domains can easily be handled by this method. The main disadvantage of the Monte-Carlo method is its computational effort but this is becoming less concern due to the revolution in the computer speed. The probability density function or the histogram of the input parameters must be known, which is not necessary in the spectral or perturbation methods. A large number of realizations is necessary in order to get meaningful statistical analysis.

Monte-Carlo approach is based mainly on generating random fields of the hydrogeological parameters to represent the heterogeneity of the formation. Then, the usual groundwater flow and/or transport equations can be solved numerically in this geometrical structure. One can assume the probability density function of the model parameters or joint probability density function for a number of parameters in the model. The assumptions of these density functions are based on some field tests and/or laboratory tests. By using a random number generator, one generates a realization for each one of these parameters. The parameter generation can be correlated or uncorrelated depending on the type of the problem. With this parameter realization a classical numerical flow or/and transport model is run and a set of results is obtained. Another random selection of the parameters is made and the model is run again, and so on. It's necessary to have a very large number of runs, and the output model results corresponding to each input is obtained which can be represented mathematically by the stochastic process $\Phi(x,\zeta_i)$. Statistical analysis of the ensemble of the output (i.e. $\Phi(x,\zeta_i)$ for $i = 1,2,......m$) can be done to get the mean, the variance, the covariance or the probability density function for each node with a location x in the grid.

2.5.3 Random Sampling in Monte-Carlo Approach

The Uni-variate random number generator was first applied to porous media flow by Warren and Price [1961] who built the first stochastic model of macroscopic flow. They extended the same approach to transport which is published in Warren and Price [1964]. The uni-variate distribution is applied to one-dimensional groundwater flow by Freeze [1975] and to one-dimensional consolidation problems by Freeze [1977]. Freeze assumed that the geostatistical parameters such as hydraulic conductivity, porosity and soil compressibility are tri-variate normal distribution. This method assumed that values of log-conductivities, porosities and compressibili-ties are not auto-correlated spatially, i.e., the parameter values of one block are still statistically independent of those of all the other blocks, on other words, the generated sequence of the parameter values are uncorrelated in space. The following sections discuss several methods for Monte-Carlo sampling.

2.5.3.1 Random Number Generation for Uncorrelated Parameter

Monte-Carlo methods rely, due to their stochastic nature, on random numbers. The simplest method for generating random sampling from a given probability density function is discussed below. The corresponding literature is immense [Johnson 1987; Tong 1990], so that only a few methods are given here. The proce-dure for generating uni-variate random sampling is as follows. Use a uniform random number generator to generate pseudo-random number in the range $\{0,1\}$. One of the famous methods is called the 'Multiplicative Congruence Method' developed by Lehmer [1951, referenced by Harbaugh and Bonham-Carter 1970; and Yevjevich, 1972]. This method is defined by the following relationships:

$$N_i = MODULO \ (A.N_{i-1} \ , \ M) \qquad (2.70)$$

$$U_i = N_i/M \qquad (2.71)$$

where, N_i is a pseudo-random integer, i is subscript of successive pseudo-random integers produced, i-1 is the immediately preceding integer, M is a large integer used as the modulus, A is an integer constant used to govern the relationship in company with M, U_i is a pseudo-random number in the range $\{0,1\}$, and '*MODULO*' notation indicates that N_i is the remainder of the division of $(A.N_{i-1})$ by M.

Most computer methods use this multiplicative congruence relationship which can be found in computer libraries or coded directly into any program. All the integers in the congruence relationship are limited by the computer word (for *IBM*

360 computer, word length = *32* bits i.e. *4* bytes).

A great deal of effort has been devoted to the identification of optimal values of the parameters in Eq.(2.70) so as to minimize errors in the results. A recommended values of these parameters as given by Naylor and others [see Harbaugh and Bonham-Carter, 1970] are, *M* is chosen as 2^{b-1}, where, *b* is the word length of the computer. For *IBM 360 M* = 2^{31} = 2,147,483,648 is a possible choice. The value of *A* is an odd integer of the form 8*t*+or-1, where, *t* is any positive integer. A value of *A* close to $2^{b/2}$ has been found to satisfy certain statistical requirements. Thus for *b* = 32, *A* = 2^{16} +3 = 65,539 is a satisfactory choice. The initial value of *N* (seed value, N_o) is an odd integer smaller than *M*.

However, the probability distribution of most parameters are not uniform. One often wishes to transform uniform distributed random numbers to other distributions. There are several methods used to transform a uniform distribution to other distributions. In brief, these methods are [Johson, 1987]: (1) Inverse of the distribution function, (2) Transformations based on special distributional relationships, and (3) Acceptance-rejection method.

(i) Inverse of The Distribution Function

The preceding section produces numbers for uniform distributions. Non-uniform distributions are easily derived from this method. It is based on transforming the probability density function $f(\alpha)$ into the cumulative frequency distribution $F(\alpha)$ by summation over each discrete class or by integration of the continuous distribution.

$$F(\alpha) = \int_{-\infty}^{\alpha} f(\alpha) \ d\alpha \qquad (2.72)$$

where, α is the parameter under study. If one assumes a uniform distribution '*U*', and the function $F(\alpha)$ is defined over the range {0,1}, then a random value of α, that will have the cumulative distribution function, can be determined from the relationship:

$$U = F(\alpha) \qquad (2.73)$$

or by taking the inverse and solving for α one can get,

$$\alpha = F^{-1}(U) \qquad (2.74)$$

which simply means that α is solved in terms of *U*. Fig.(2.4) shows how this method works graphically.

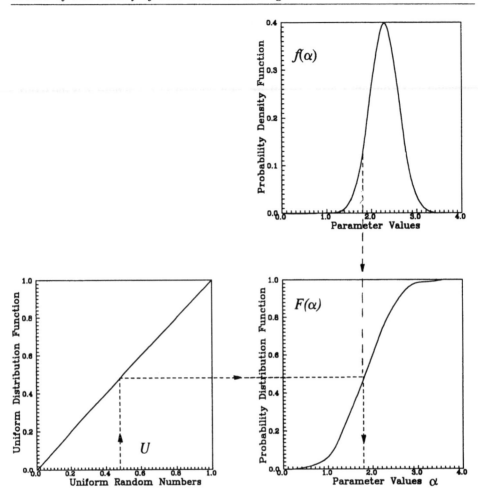

Fig.(2.4) Transformation of Random Numbers of The Uniform Distribution Function 'U' to Random Numbers of any other Distribution Function F(α).

(ii) Transformation Method

A transformation other than $F^{-1}(U)$ can be used for Monte-Carlo sampling. Two applications of this method will be given here, which are commonly used for generating hydrogeological parameters in groundwater flow modelling.

Random Number Generator for Normal Distribution
Generation of normally distributed random number, ε, can be obtained by the following formula [Naylor et al., 1966 referenced by Harbaugh et al., 1970],

$$\varepsilon = \frac{\sum\limits_{i=1}^{m} U_i - m/2}{\sqrt{m/12}} \qquad (2.75)$$

with mean ($\mu=0$) and unit standard deviation ($\sigma=1$), U_i is the i-th element of a sequence of random numbers from a uniform distribution in the range $\{0,1\}$, and m is the number of U_i to be used.

This method is based on the central limit theorem which states that 'Observations which are the sum of many independently operating processes tend to be normally distributed as the number of effects becomes large' [Mood and Graybill, 1963]. By summing a number m of uniform random numbers, a normal random number is obtained. If m is 12, a normal distribution with tails truncated at six times standard deviation is produced and the general equation, Eq.(2.75) becomes:

$$\varepsilon = \sum\limits_{i=1}^{i=12} U_i - 6 \qquad (2.76)$$

To transform ε from standard normal distribution form $N(0,1)$ to any normal distribution $N(\mu_\alpha, \sigma_\alpha)$ is simply by multiplying ε by σ_α and adding μ_α i.e.

$$\alpha = \mu_\alpha + \varepsilon \, \sigma_\alpha \qquad (2.77)$$

An alternative and direct method for generating normally distributed random numbers has been referred to by several authors [Box and Muller, 1958 referenced Harbaugh et al., 1970; Yevjevich, 1972; Johnson, 1987; Koonin and Meredith, 1990]. This method gives a higher degree of accuracy than the inverse method. The transformation formulas used by this method are

$$\varepsilon_1 = (-2 \, Ln \, U_1)^{1/2} \, \cos(2\pi U_2) \qquad (2.78)$$

$$\varepsilon_2 = (-2 \, Ln \, U_1)^{1/2} \, \sin(2\pi U_2) \qquad (2.79)$$

where, U_1 and U_2 are independent random numbers distributed in the range $\{0,1\}$, and ε_1 and ε_2 are independent standard normally distributed random numbers with zero mean ($\mu=0$) and unit standard deviation ($\sigma=1$).

Equations similar to Eq.(2.77) can be written as follows,

$$\alpha_1 = \mu_\alpha + \varepsilon_1\, \sigma_\alpha$$
$$\alpha_2 = \mu_\alpha + \varepsilon_2\, \sigma_\alpha \qquad\qquad (2.80)$$

where, α_1 and α_2 are independent normally distributed random numbers $N(\mu_\alpha, \sigma_\alpha)$.

Random Number Generator for Log-Normal Distribution
For a log-normal distribution generator a transformation method can be used in the form,

$$y = e^\alpha \qquad\qquad (2.81)$$

This means that generation of α as $N(\mu_\alpha, \sigma_\alpha)$ gives a generation of y which is log-normal distribution $LN(\mu_y, \sigma_y)$. Generation of log-normally distributed sets of random numbers involves the methods mentioned for normally distributed random numbers and the transformation formula Eq.(2.81).

(iii) Acceptance-Rejection (AR) Method

This is another approach for generating random numbers drawn from a specific probability density functions. The *AR* method provides a general procedure for generating random varieties with any distribution whose probability density $f(\alpha)$ is continuous and bounded within a finite region. For more details about the method see Gottfried [1984]. Although the *AR* method can be used with many different probability density functions, it is relatively inefficient. Thus the *AR* method is generally used only if there is no other alternative.

2.5.3.2 Random Number Generator for Auto-Correlated Parameter

A parameter value is said to be auto-correlated, if the value of the parameter in some point is related closely to the value in the nearby points. If a geostatistical parameter is highly auto-correlated, all of the neighbouring points will have approximately the same value. In contrast, if a geostatistical parameter is poorly auto-correlated, neighbouring points will have little relation to one another. Fig.(2.5) shows typical examples for the difference between highly auto-correlated and poorly auto-correlated series of data.

There are several methods in the literature used for generating random sequences with some auto-correlation structure. The most famous methods will be described in this section. Attention will be given to the Gaussian random fields, where most of the geostatistical parameters exhibit variability which follows a normal or log-normal probability density function. These methods can be classified as: (1) Direct methods or matrix methods, such as, multi-variate normal distribution and nearest

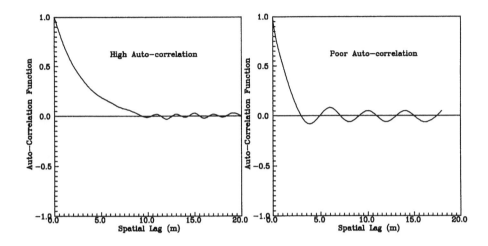

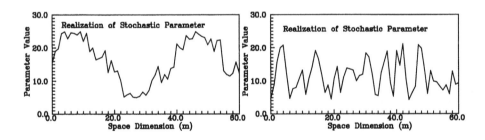

Fig.(2.5) Some Space Series and Their Auto-Correlation Functions.

neighbour method. (2) Indirect or transformation methods, such as, turning bands algorithm.

(i) Multi-Variate Normal Random Number Generator

In the uni-variate case as discussed in the preceding sections, only one random variable was considered. The multi-variate case given in this section is a natural extension of the uni-variate case, i.e., more than one random variable or what is called random vector will be discussed. A p-dimensional random vector $Z = \{Z_1, Z_2, Z_3, \ldots\ldots Z_p\}^T$ is defined to have a multi-variate normal distribution if and only if the p-components of Z have a uni-variate normal distribution. The distribution

39

of $\mathbf{Z}$ is denoted $N_p(\mu,C)$. The multi-variate normal density function $N_p(\mu,C)$, p-dimensional normal random variate, is given by Mood and Graybill [1963],

$$f(\mathbf{Z}) = \frac{1}{|C|^{1/2}(2\pi)^{p/2}} \exp\left[-\frac{1}{2}(\mathbf{Z}-\mu)^T C^{-1}(\mathbf{Z}-\mu)\right] \qquad (2.82)$$

where, $N_p(\mu,C)$ is a multi-variate normal distribution with mean μ, and covariance matrix C, p is the number of parameters (nodes of the model), $\mathbf{Z} = \{Z_1, Z_2,........, Z_p\}^T$, p-dimensional random vector, $(p*1)$, $\mu = \{\mu_1, \mu_2,.........._p\}^T$, p-dimensional mean values vector, $(p*1)$, T is superscript transpose operation of a matrix, -1 superscript is inverse operation of a matrix, C is a $p{\times}p$ covariance matrix given by,

$$C = \begin{bmatrix} \sigma^2_{Z_1} & Cov(Z_1,Z_2) & .... & Cov(Z_1,Z_p) \\ Cov(Z_2,Z_1) & \sigma^2_{Z_2} & .... & Cov(Z_2,Z_p) \\ . & & & \\ . & & & \\ Cov(Z_p,Z_1) & Cov(Z_p,Z_2) & ..... & \sigma^2_{Z_p} \end{bmatrix} \qquad (2.83)$$

$|C|$ is the determinant of the covariance matrix C, and σ^2_{Zi} is the variance of the random parameter Z_i.

All the diagonal elements of the covariance matrix are the variances of the individual random variables and the off-diagonal components are covariances of two random variables Z_i, Z_j where, $i = 1,.....p$, and $j = 1,.....p$. The covariance matrix can be defined as the matrix which lists the correlations between the parameters. It should be noted that all the covariance matrices are symmetric. This property allows us to apply the theory of symmetric matrices. Sometimes it is convenient to normalize covariance matrices by converting the individual covariance in terms of the correlation coefficient ρ_{ij} as,

$$\rho_{ij} = \frac{Cov(Z_i, Z_j)}{\sigma_{Z_i}\,\sigma_{Z_j}} \qquad (2.84)$$

where, $Cov(Z_i,Z_j)$ is the covariance between Z_i and Z_j.

The generation of varieties with multi-variate distributions is more than with uni-variate distributions. The obvious difference is that in the generation of multi-variate distributions the dependencies (correlations) among the all components of the random vector must be handled.

The general technique for the generation of multi-variate distributions is called

'*Conditional Distribution Approach*'. A brief description of this approach is given here. For more details about this method reference is made to Johnson [1987].

Let $\mathbf{Z} = \{Z_1, Z_2, Z_3, \ldots\ldots Z_p\}^T$ be the p-dimensional random vector of interest. The conditional distribution approach involves the following steps: (1) generate $Z_1 = z_1$ from the marginal distribution of Z_1 (uni-variate distribution of Z_1); (2) generate $Z_2 = z_2$ from the conditional distribution of Z_2 given $Z_1 = z_1$; (3) generate $Z_3 = z_3$ from the conditional distribution of Z_3 given $Z_1 = z_1$ and $Z_2 = z_2$........... and so forth through the p steps.

In certain statistical applications the covariance matrix gives the dependencies between the components of the random vectors. In case of random fields the same philosophy is applied. Each node in the field is considered as a component of a random vector which contains all the nodes in the field. The dependencies between the components are described by the auto-covariance function of the field.

The method of multi-variate random number generation with a given covariance matrix of the system has been developed by [Iman, 1980, referenced by Peck, A. et al., editors, 1988]. It has been applied to groundwater flow by Townley [1983, 1984, 1985, 1988]. The algorithm for generating random fields with a given covariance structure based on the covariance matrix of the system is as follows [Neuman, 1984]:

1) Build the covariance matrix C of the system. The elements of C are denoted by,

$$c_{ij} = Cov \ (Z_i, Z_j) \tag{2.85}$$

and if $i=j$ the covariances becomes the variances.

If the random field is assumed to be statistically homogeneous (stationary), i.e., the mean is constant, the standard deviation is constant, and the covariance depends only on the distance vector between the points in the domain, then one can write the elements of the covariance matrix as,

$$c_{ij} = \sigma^2_z \ \rho(s_{ij}) \tag{2.86}$$

where, σ^2_z is the variance of the process Z, $\rho(s_{ij})$ is the auto-correlation function, and s_{ij} is the distance vector between point i and point j. Substitution of the covariance equation, Eq.(2.86), into the elements of covariance matrix and dividing the matrix by σ^2_z, one can get the so-called correlation matrix, which takes the general form,

$$R = \frac{1}{\sigma^2_z}(c_{ij}) = \begin{bmatrix} 1. & \rho_{12} & \cdots\cdots & \rho_{1p} \\ \rho_{21} & 1. & \cdots\cdots & \rho_{2p} \\ \cdot & & & \\ \cdot & & & \\ \cdot & & & \\ \rho_{pl} & \rho_{pl} & \cdots\cdots & 1. \end{bmatrix} \qquad (2.87)$$

where, $\rho_{ij} = c_{ij} /\sigma^2_z$ is the correlation coefficient between point i and point j. All the diagonal elements of the correlation matrix R are equal to one, which means that the correlation between the point and itself is perfect (complete correlation). If $\rho_{ij} = 0$, this means no correlation between i and j. All the off- diagonal elements are called correlation coefficients and they are less than one. Addition-ally, the correlation matrix is symmetric, i.e., $\rho_{ij} = \rho_{ji}$. According to the stationarity assumption, the value of $\rho_{12} = \rho_{23} = \ldots\ldots\ldots= \rho_{p-1p}$, the value of $\rho_{13} = \rho_{24} =\ldots\ldots\ldots=$ ρ_{p-2p}, and so on.

The correlation matrix can be calculated for a certain random field given its auto-correlation function as follows. For pairs of values Z_i and Z_j with $i = 1,...p$ and $j = 1,...p$ determine $(x_i - x_j)$ and $(y_i - y_j)$ where, (x_i, y_i) are the coordinates of point Z_i and point (x_j, y_j) are the corresponding coordinates for point Z_j. The distance s_{ij} between two points is,

$$s_{ij} = \sqrt{(x_i - x_j)^2 + (y_i - y_j)^2} \qquad (2.88)$$

If the $\rho(s_{ij})$ is isotropic then one can get circular contours, which means that, it depends on the magnitude of s_{ij} only. On the other hands, if the $\rho(s_{ij})$ is anisotropic then one can get elliptic contours, which means that, it depends not only on the magnitude of s_{ij} but also on the orientation of the vector s_{ij}. For a known auto-correlation function one can determine all relevant values of ρ_{ij}.

2) One has to decompose the covariance matrix by the Cholesky factorization method Westlake [1968] into

$$C = L\,U \qquad (2.89)$$

where, L is a unique lower triangular matrix, U is a unique upper triangular matrix, and U is L^T which means that U is the transpose of L.
Cholesky method is sometimes called '*Square-Root*' method which is used only for the decomposition of symmetric matrices. For more details about Cholesky

method see Westlake [1968]. In brief, the elements of the lower triangler matrix are given by,

$$l_{i1} = \frac{c_{i1}}{\sqrt{c_{11}}} \quad , \quad 1 \leq i \leq p \tag{2.90}$$

$$l_{ii} = \left[c_{ii} - \sum_{k=1}^{i-1} l_{ik}^2 \right]^{1/2} , \quad 1 \leq i \leq p$$

$$l_{ij} = \frac{1}{l_{jj}} \left[c_{ij} - \sum_{k=1}^{j-1} l_{ik} l_{jk} \right] , \quad 1 < j < i \leq p \tag{2.91}$$

$$l_{ij} = 0 , \qquad\qquad i < j \leq p$$

3) Generation of normally distributed p-dimensional sequence of independent random numbers with zero mean and unit standard deviation $N(0,1)$ which can be expressed as, $\varepsilon = \{\varepsilon_1, \varepsilon_2, ..., \varepsilon_p\}^T$ where, ε is vector of normally distributed random numbers, and ε_i is the i-th random number drawn from $N(0,1)$. One of the uni-variate random number generator methods which were discussed in the previous sections can be used for the generation of the normal random vector ε.

4) Multiplication of the independent random vector ε by the triangular matrix U to get a vector of correlated random numbers. This vector can be expressed by matrix multiplication convention as,

$$X = U \varepsilon \tag{2.92}$$

where, X is a vector of multi-variate normal random $N_p(0,I)$, 0 is zero mean vector ($p*1$), and I is the identity matrix ($p*p$). The desired random field is then obtained by

$$Z = \mu + X \tag{2.93}$$

The random process Z is the one for which the samples $Z_1, Z_2, ..., Z_p$ are jointly distributed according to a multi-variate probability density function that is given by Eq.(2.82). In the Monte-Carlo approach, the matrix decomposition by the Cholesky method is performed only once, then a large number of correlated fields can be generated at very low computer cost because this only involves the product of a vector of random numbers by the decomposed matrix.

(ii) Nearest Neighbour Method

Another method which is used for generating a correlated random field is called *'Nearest Neighbour Method'*. The general framework required for this method was presented by Whittle [1954]. This method is sometimes called Whittle Model. The idea of this method is based on generating an independent random field with one of the methods described in the previous sections. Then, the value of a parameter at a given node is replaced by weighted average of previous random values at the given node and a few surrounding ones. This model was first applied to one-dimensional steady state groundwater flow in a bounded domain by Smith and Freeze [1979a] and then they also extended the same approach to the two-dimensional steady state groundwater flow in [1979b]. This method can be expanded to three-dimensions. A general description of this spatial model used by Smith and Freeze can be written in a generalized form proposed by Whittle [1954],

$$Z_i = \sum_{j \ne i} W_{ij} Z_j + \varepsilon_i \qquad (2.94)$$

where, Z_i is random variable satisfying the nearest neighbour relation, ε_i is uncorrelated normal random number with $E(\varepsilon_i) = 0$, and $Var(\varepsilon_i) = \sigma_i^2$, $i=1,2,.........p$, and W_{ij} are weighting coefficients.

In case of the anisotropic first-order auto-regressive scheme, Eq.(2.94) can be rewritten according to Smith and Freeze [1979b] as,

$$Z_{ij} = \alpha_x(Z_{i-1j} + Z_{i+1j}) + \alpha_y(Z_{ij-1} + Z_{ij+1}) + \varepsilon_{ij} \qquad (2.95)$$

where, α_x is an auto-regressive parameter expressing the degree of dependence of Z_{ij} on its two neighbouring values in the horizontal direction, Z_{i-1j} and Z_{i+1j}, ($|\alpha_x| < 1$), and α_y is an auto-regressive parameter expressing the degree of dependence of Z_{ij} on its two neighbouring values in the vertical direction, Z_{ij-1} and Z_{ij+1}, ($|\alpha_y| < 1$).

Eq.(2.95) should be applied with modifications at the boundaries of the domain of interest where necessary, because the blocks do not extend across the boundaries i.e. the equation will be truncated near the boundary. Eq.(2.95) can be recanted in matrix notation as,

$$\mathbf{Z} = \mathbf{W} \mathbf{Z} + \varepsilon \qquad (2.96)$$

where, matrix $\mathbf{W}$ is called the $p*p$ connectivity matrix, or the $p*p$ spatial lag operator of scaled weights, w_{kl}. The elements of the connectivity matrix w_{kl} are defined as,

$$w_{kl} = \frac{w^*_{kl}}{N} \qquad (2.97)$$

where, $k = 1,2,...p$, $l = 1,2,...p$, and $k \neq l$, $w^*_{kl} = \alpha_x$ if the blocks k and l are contiguous in the x-direction, $w^*_{kl} = \alpha_y$ if the blocks k and l are contiguous in the y-direction, and $w^*_{kl} = 0$ otherwise, i.e., if $k = l$, or if blocks k and l are not contiguous, and N is the total number of blocks surrounding block k, i.e., $N=4$ if block k is located inside the domain that is being modeled, $N=3$ if block k is located on the boundary of the domain, and $N=2$ if block k is located at a corner of the domain.

The scaling of N is required to ensure stationarity in the generated sequence of Z_{ij} values near the domain boundaries because Eq.(2.96) is truncated near the boundaries.

The Z sequence has a mean μ_Z and standard deviation σ_Z. Similarly, the ε has mean μ_ε and standard deviation σ_ε. In order to simulate the predetermined standard deviation σ_Z, one should start from a random sequence ε with a standard deviation of one. This vector can be premultiplied by an appropriate factor to yield the desired value of σ_Z. This factor is denoted by η. Eq.(2.96) becomes,

$$\mathbf{Z} = \mathbf{W}\,\mathbf{Z} + \eta\,\varepsilon \qquad (2.98)$$

Eq.(2.98) can be solved for Z sequence as follows

$$\mathbf{Z} = (\mathbf{I} - \mathbf{W})^{-1}\,\eta\,\varepsilon \qquad (2.99)$$

To determine the simulation parameter η, one can follow the procedure of Smith and Freeze [1979] or the procedure of [Baker, 1984] as given in brief here. The auto-correlation matrix of the process $\mathbf{Z}$ can be written as

$$\mathbf{R} = \frac{E(\mathbf{Z}\,\mathbf{Z}^T)}{\sigma_Z^2} \qquad (2.100)$$

Substitution of Eq.(2.99) into Eq.(2.100) and with some algebraic manipulations, one gets

$$\mathbf{R} = \frac{1}{\sigma_Z^2}\,\mathbf{V}\,\eta^2 . \sigma_\varepsilon^2 \qquad (2.101)$$

where, $\mathbf{V} = (\mathbf{I} - \mathbf{W})^{-1}.((\mathbf{I} - \mathbf{W})^{-1})^T = ((\mathbf{I} - \mathbf{W}).(\mathbf{I} - \mathbf{W})^T)^{-1}$
Let σ_ε be equal to one in Eq.(2.101), then

$$R = \frac{1}{\sigma_Z^2} V \eta^2 \tag{2.102}$$

All the diagonal elements of R are equal to unity by definition, hence, by taking the trace (the sum of the elements on the main diagonal of the square matrix) and solving the resultant scalar equation for η gives,

$$\eta = \frac{\sigma_Z}{\sqrt{V_m}} \tag{2.103}$$

where, $V_m = tr \ V/p$, and the symbol '*tr*' is the trace of the matrix, which is given by, $tr \ V = \sum v_{ii}$, $i = 1,\ldots\ldots,p$. Substituting Eq.(2.103) into Eq.(2.99), the final nearest neighbour generator is obtained.

$$Z = (I - W)^{-1} \frac{\sigma_Z}{\sqrt{V_m}} \varepsilon \tag{2.104}$$

where, the Z sequences have a mean of zero. By adding the constant μ_Z to each element of Z the system of equations for the nearest neighbour model can be written as

$$Z = \mu_Z + (I - W)^{-1} . \frac{\sigma_Z}{\sqrt{V_m}} \varepsilon \tag{2.105}$$

The analysis of the covariance function describing the generated random field with first-order dependence is approximately an exponential decay function Smith and Freeze [1979]. By adjusting the number of neighbours (higher-order nearest neighbour models), the weighting coefficients and the variance of the initial random independent parameters, it is possible to approximately fit any given real covariance function as observed on the data.

The advantage of this technique can be seen in Eq.(2.105). At the beginning of any simulation the matrix $(I - W)$ must be inverted only once. For each realization of the process Z, the inverted matrix $(I - W)^{-1}$ is simply multiplied by the generated random vector $\eta\varepsilon$. The drawback of this method is computing the inverse matrix.

(iii) Turning Bands Method (TBM)

The turning bands method is one of the techniques which is designed to generate a realization of stationary, correlated, and multidimensional Gaussian random field from a normal distribution with zero mean and a specified covariance structure. The *TBM* was first proposed by Matheron [1973] and applied by the Ecole des Mines de Paris [e.g. Journel, 1974; and Delhomme, 1979].

The *TBM* is based on the theory of random fields (multidimensional stochastic process). Its basic concept is to transform a multidimensional simulation into the sum of a series of equivalent uni-dimensional simulations [Mantoglou and Wilson, 1982]. The basic idea of the algorithm in brief is, generating two- and three-dimensional fields by subsequent projection and combining values found from a series of one-dimensional simulations along several lines radiating outward from an arbitrary origin in space. This procedure yields discrete values or realizations of the random field. This method has been widely used in porous flow and transport studies.

TBM is a repetition of a two steps procedure. First, a realization of a random process with a prescribed auto-covariance function and zero mean is generated on one line. The Cholesky decomposition method can be used (but with much smaller correlation matrix dimensions) or by auto-regression methods, like nearest neighbour. Second, orthogonal projection of the generated line process to each point in the simulated two- or three-dimensional random field. The two steps are repeated for a given number of lines and then a final value is assigned to each grid point in the field by taking a weighted average over the total number of lines.

There are two main approaches for generating the one-dimensional line process in *TBM*. The first one is space domain approach which was first proposed by Matheron and applied by Journal and Huijbregts [1978] and Delhomme [1979]. This approach can handle only particular forms of auto-correlation functions. The second one is the spectral (frequency) domain approach *STBM*. This approach has been implemented by Mantoglou and Wilson [1982]. It is a more general approach which can handle a wide variety of two-dimensional processes, multivariate (cross-correlated) processes, as well as spatial averaged (integrated) processes.

(a) Theoretical Background of TBM:

Let $Z_i(u)$, $i = 1,....L$ a set of N independent realizations of a one-dimensional, second order stationarity stochastic process on a line u with an auto-correlation function $\rho_1(u_o)$, where u_o is the spatial lag on the line. Then the values given by the relation,

$$Z_s(x,y,z) = \frac{1}{\sqrt{L}} \sum_{i=1}^{L} Z_i(u) \qquad (2.106)$$

is a realization of a two- or three-dimensional process. The subscript s represents the term *'simulated'* or *'synthetic'*. The field generated by Eq.(2.106) has zero mean as well. The relation between the auto-correlation function on the line process $\rho_1(u_o)$ and the auto-correlation function in the three-dimensional random field $\rho(u_o)$ is given by [for the derivation, refer to Mantoglou and Wilson, 1982; and Mantoglou, 1987],

$$\rho_1(u_o) = \frac{d}{du_o}[u_o\, \rho(u_o)] \qquad (2.107)$$

and for two-dimensional field the relationship becomes,

$$\int_0^s \frac{\rho_1(u_o)du_o}{\sqrt{(s^2 - u_o^2)}} = \frac{\pi}{2}\, \rho(s) \qquad (2.108)$$

where, s is the spatial lag in two-dimensional field.

It can be observed from Eq.(2.108) that, it is not easy to obtain $\rho_1(u_o)$ directly as a function of $\rho(s)$. Sironvalle [1980] wrote [referenced by Baker, 1984] 'This integral equation is too difficult to solve, so we discard this (the *TBM*) method.' Therefore, Mantoglou and Wilson [1982] used a spectral method to generate uni-dimensional process along the lines for different types of auto-correlation functions. In the following section this method will be given more attention.

(b) Spectral Turning Bands Method (STBM):

Generation of 2-D random fields by *TBM* needs the solution of the integral equation given by Eq.(2.108). This integral equation cannot be directly expressed as a function of $\rho(s)$. Particular solutions can be found for certain two-dimensional auto-correlation functions [Mantoglou and Wilson, 1982]. To circumvent this difficulty, an expression for the spectral density function of the one-dimensional processes as a function of the radial spectral density function of the two-dimensional process is used. This expression is given in Fourier space by,

$$S_1(\omega) = \frac{\sigma_z^2}{2} \, S(\omega) \qquad (2.109)$$

This means, that the spectral density function of the uni-dimensional process $S_1(\omega)$ along the turning bands lines is given by one half of the radial spectral density function $S(\omega)$ of the two-dimensional process multiplied by the variance of the two-dimensional process.

Steps used in implementing the *STBM* generator in simulating a two-dimensional random field are given below:

(1) Generation of One-dimensional Uni-variate Process on The Turning Bands Line:

In the literature, there are two main techniques used for generation of the line process. The first, is called the *'Fast Fourier Transform'* (*FFT*) algorithm, which can be used to construct a complex process $X(u) = Z(u) + i \, Y(u)$ given by Tompson, et al. [1989],

$$X(u) = \int_{all\omega} e^{i\omega u} \, dW(\omega) \approx \sum_{all\omega} e^{i\omega_j u} \, dW(\omega_j) \qquad (2.110)$$

where, in the approximate form, X is the sum of a complex series of sinusoidal functions of varying wavelength, each magnified by complex random amplitude with zero mean $dW(\omega_j)$, $\omega_j = j.\Delta\omega$.

The second is the *'Standard Fourier Integration'* method. The real part of the complex process $X(u)$ is given by

$$Re \; X(u) = Z(u) = \int_{all\omega} |dW(\omega)| \; \cos(\omega \; u + \phi_\omega) \qquad (2.111)$$

can be used to develop a straightforward discrete approximation using positive frequencies,

$$Z_i(u) = \sum_{j=1}^{M} |dW(\omega_j)| \; \cos(\omega_j \; u + \phi_j) \qquad (2.112)$$

where, ϕ_j represents independent random angles which is uniformly distributed between 0 and 2π, M is the number of harmonics used in the calculations, $\omega_j = (j - .5) \, \Delta\omega$, $j = 1,2,...M$, and $\Delta\omega$ is the discretized frequency which is given by

ω_{max}/M, and ω_{max} is the maximum frequency used in the calculations. The magnitude $|dW(\omega_j)|$ is taken to be deterministic from the spectrum as,

$$|dW(\omega_j)| = \left[4\ S_1(\omega_j).\Delta\omega\right]^{1/2} \tag{2.113}$$

where, $S_1(\omega_j)$ is the spectral density function of the real process $Z(u)$ on the line.

$S_1(\omega)$ is assumed to be insignificant outside the region $[-\omega_{max},+\omega_{max}]$. Substitution of Eq.(2.113) into Eq.(2.112) gives the generator of the uni-dimensional process on line i as,

$$Z_i(u) = 2\sum_{j=1}^{M}\left[S_1(\omega_j).\Delta\omega\right]^{1/2}\cos(\omega_j\ u + \phi_j) \tag{2.114}$$

This is the classical method proposed by Rice [1954]. The form of Eq.(2.114) is slightly modified by Shinozuka and Jan [1972] to give,

$$Z_i(u) = 2\sum_{j=1}^{M}\left[S_1(\omega_j).\Delta\omega\right]^{1/2}\cos(\acute{\omega}_j\ u + \phi_j) \tag{2.115}$$

where, $\acute{\omega}_j = \omega_j + \delta\omega$

The frequency $\delta\omega$ is a small random frequency added here in order to avoid periodicities. $\delta\omega$ is uniformly distributed between $-\Delta\acute{\omega}/2$ and $\Delta\acute{\omega}/2$, where, $\Delta\acute{\omega}$ is a small frequency, $\Delta\acute{\omega}<<\Delta\omega$. $\Delta\acute{\omega}$ is taken equal to $\Delta\omega/20$ according to Shinozuka and Jan [1972].

This approach involves a discrete summation over frequency increments $\Delta\omega$ up to a maximum cutoff frequency of $\omega_{max} = M\ \Delta\omega$. The method involves a larger computational effort than *FFT*, but it is more flexible in the choice of the parameters, M, $\Delta\omega$, Δu, ω_{max} and u_{max} [Tompson, et al., 1989].

(2) Distribution of The Turning Bands Lines and The Number of Lines:

Theory of TBM is based on an infinite number of lines. The lines are assumed to be randomly oriented, as taken from a uniform distribution on a unit circle in 2-D space or sphere in 3-D space. It has been shown by Mantoglou and Wilson [1982] that by spacing the lines evenly on the unit circle or sphere, with prescribed directions, the simulated correlation function converges much faster to the theoretical function. Mantoglou and Wilson show that a number of 8 -16 lines is generally a satisfactory choice in isotropic auto-correlation function. In anisotropic case, however, a larger number of lines might be required.

(3) Spectral Discretization:

A realization of the line process $Z_i(u_n)$ at point n on line i is constructed from a discrete integration of a series of random components over all the frequency domain. The frequency discretization $\Delta\omega$ must be kept small enough to ensure a sufficient degree of accuracy, while the number of harmonics M must be kept large enough to account for the contributions of the spectral tail at $\omega_{max} = M \Delta\omega$. Mantoglou and Wilson [1982] considered the case where M is varying between 50 and 100 and ω_{max} is 40 times the correlation length of isotropic auto-correlation function in both cases. They found that while the accuracy for $M = 50$ is poor at large distances, the accuracy shown for $M = 100$ improves rapidly.

(4) Physical Discretization:

The physical increment Δu used along the line should be chosen less than the domain discretization $\Delta x, \Delta y$ as a pragmatic rule of thumb in order to avoid some numerical problems [Mantoglou and Wilson, 1982].

(5) Length of The Turning Bands Lines:

The minimum length of the line is determined by the orientation of the line and the size of the simulation domain.

(6) Generation of The Simulated Field:

The construction of a simulated field $Z_s(x,y,z)$ will include the selection of a finite number of lines L and their orientation, the discrete simulation of one-dimensional process $Z_i(u)$ on each line, the subsequent projection of these values onto all simulation points x in the domain, and the division of each sum at each point by $L^{1/2}$ to yield $Z_s(x,y,z)$.

2.6 Conditional Simulations of Stationary Random Fields upon Measurements of Field Parameters

All the stochastic methods presented thus far are used for unconditional simulation of stationary random fields. These methods reproduce, the first and second moments or the *PDF* of the simulated field. From a practical point of view, it is desirable that the random fields not only reproduce the spatial structure of the field but also honour the measured data and their locations. This requires an implementation of some kind of conditioning, so that the generated realizations are constrained to the available field measurements. One of the advantages of conditioning is to reduce uncertainty in the simulated fields. The uncertainty will be reduced significantly (perhaps to zero in the absence of measurements errors) at the sampled locations and it will be reduced in the vicinity of the sampled locations. The conditional simulation can be achieved by spatial estimation approaches such as Kriging. Kriging is known as the best linear, unbiased estimator (*BLUE*). Therefore, it has advantages over the conventional spatial interpolator techniques (e.g. least square method, weighted residuals method, etc. [see for more details Christakos, G., 1992]) in providing not only the estimates of point values, but also the variance of the corresponding errors of estimation (uncertainty associated with the estimate). One drawback of the Kriging method is that the estimated field is smoother than the real data. Therefore using conditional simulation with Kriging reproduces closely the true variability of the field and honours the measurements. The methods of conditioning can be broadly divided into:

(*i*)'*Direct Approaches or Matrical Methods*': these approaches draw the realizations directly from the sub-ensemble of conditional realizations. The famous algorithms are Gaussian conditional simulations via the *LU* triangular decomposition of the autocovariance matrix (see section 2.5.3.2 *i*) and it is presented in more details by Davis [1987]. This method has been applied to subsurface hydrology by Neuman [1984]. The limitation of this method is that it is restricted in the size of the grid it can handle. Another method is called sequential simulation algorithm suggested by Journal and Huijbreghts [1981]. This method works by calculating the conditional distribution of grid point values given the data values at the sampled locations and by assuming a multivariate normal distribution of the given data.

(*ii*)'*Indirect Approaches*': these approaches are considered as two steps. Firstly an unconditional realization is generated. Secondly, a modification is used to honour the data at their locations. This method works simply by adding a correlated error obtained from the simulation to a kriged map from the sampled data. In brief, the procedure is the following:

(1) A kriged map is generated from the field data with the sampled locations which will be smoother than reality.

(2) An unconditional simulated field is generated by *TBM* from the data which reproduces the spatial structure of the underlying random function.

(3) Allocation of the unconditional values (pseudo measurements) at the sites of measurements is done on the simulated map in step 2.

(4) Another kriged map is generated from the pseudo measurements.

(5) A pseudo error is calculated by subtracting the kriged map in step 4 from the unconditional simulation in step 2.

(6) The conditional simulation map is generated by adding the pseudo error in step 5 to the kriged map in step 1. So,

$$Z_{cs} = Z_{kd} + (Z_{us} - Z_{kus}) \tag{2.116}$$

where, Z_{cs} is the required conditional simulation, Z_{kd} is the kriged map from the real data, Z_{us} is the unconditional simulation, Z_{kus} is the kriged map with the pseudo measurements.

Conditional simulation has been implemented analytically in flow models by [Dagan, 1982a] under the assumption of multi-Gaussian distribution where conditional probabilities can be expressed analytically or by Kriging equations.

2.7 Sources of Errors in Simulation of Random Fields

There are two main types of errors in the simulation of random fields. The first is errors due to using a simple model to match reality. This kind of errors includes the assumption on isotropy, stationarity, constant means and errors in the estimation of the covariance model from available data. A collection of sufficient field and laboratory data either hard or soft is required to minimize these errors and to provide a good estimation of the required parameters (mean, variance, auto-covariance function, etc.). The second is related to the errors within the simulations. These are errors within single realizations caused by discretization of the domain and the generation equations. These errors can also be minimized by fulfilling certain numerical rules given by each method. There are other sorts of errors within the simulation which are called estimation errors of multiple realizations introduced by the use of finite number of realization to estimate ensemble statistics. These errors can also be minimized by choosing a sufficient number of realizations.

2.8 Evaluation of The Different Techniques

In this section, a comparison of the techniques has been worked out and a discussion of the advantages and disadvantages of these methods are tabulated in Table (2.1) and Table (2.2). Table (2.1) displays a comparison between analytical and numerical approaches used for solving stochastic differential equations. Table (2.2) displays another comparison between the different techniques applied for generation of random fields which are used as an input for numerical simulation approach (Monte-Carlo). Fig.(2.6) shows three single realizations of log-hydraulic conductivity generated in a domain of dimensions 15×15 m by Nearest Neighbour Generator, *NNG*, Multi-Variate Generator, *MVG*, and Turning Bands Generator, *TBG*, worked by Elfeki [1993]. The domain is discretized by 1 m in both horizontal and vertical directions. The parameters used for the simulations are $\langle K \rangle = 1$ m/day and $\sigma_K = 2$ m/day and the corresponding logarithmic transform are $\langle Y \rangle = -0.8$ and $\sigma_Y = 1.3$. The

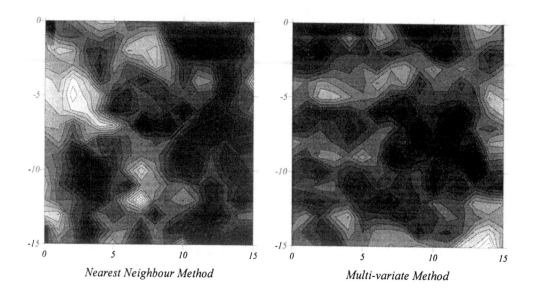

Nearest Neighbour Method   *Multi-variate Method*

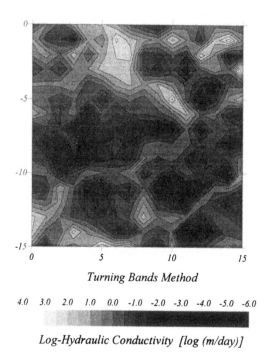

Turning Bands Method

4.0   3.0   2.0   1.0   0.0   -1.0   -2.0   -3.0   -4.0   -5.0   -6.0

Log-Hydraulic Conductivity [log (m/day)]

Fig.(2.6) Three Single Realizations Generated by Different Methods.

54

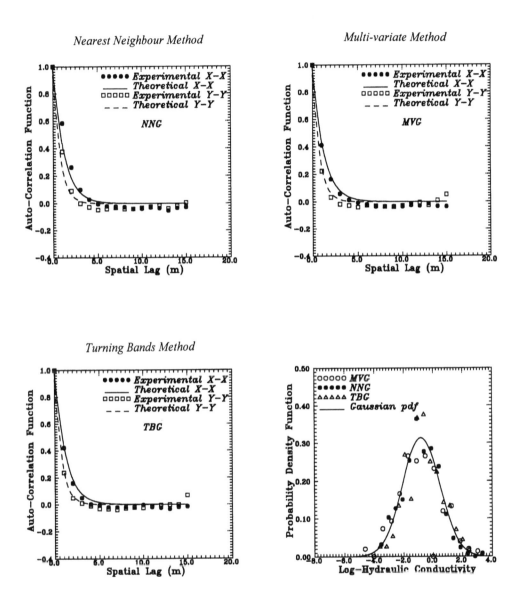

Fig.(2.7) Auto-correlation and PDF of the Generated Fields in Fig.(2.6).

Table(2.1) Comparison between Analytical and Numerical Simulation Methods.

Solution Method Item of Comparison	Analytical Methods	Monte Carlo Simulation Methods
(1) solution	defined over a continuum.	defined over a grid.
(2) stationarity of the variables	input and output variables should be stationary.	no need for stationarity assumption.
(3) probability distribution of input variables	no need to define PDF of the input variable in some applications.	the PDF of the input variables must be known.
(4) handling variability	limited to small variability.	not limited to small variability.
(5) finite versus infinite domain	derived over infinite domain.	it handles only finite but very large domains.
(6) spatial structure of the variability	simple forms of auto-covariance models.	simple and compound (nested) forms of auto-covariances.
(7) linearity versus non-linearity	based on linearized theories or weakly-nonlinearity.	it can address both cases.

(8) the outcome of the method	closed form solution of moments of dependent variable as a function of moments of independent variable. (limited only for the first two moments)	numerical values used to calculate moments of the independent variables. As the number of realizations increases, the results tend asymptotically to the exact solution. (One can calculate the complete PDF).
(9) sources of errors	errors are introduced because of the number of simplifying assumptions such as, the form of mean and covariance function, the geometry of the domain and the boundary conditions.	sampling errors are introduced because of finite number of realizations, and discretization errors are introduced because of approximation of the governing equations.
(10) time and computer effort	limited (to calculate the values).	time consuming.
(11) performing conditioning to field measurements	difficult	easy
(12) handling more than one stochastic variable	if it is possible, it is too difficult.	it is easy to handle more than one variable.

Table(2.2) Comparison between Different Numerical Simulation Methods.

Method of Simulation	Direct Methods (Matrix Methods)		Indirect Methods
Item of Comparison	Multi-Variate (MVG)	Nearest Neighbour (NNG)	Turning Bands (TBG)
(1). field data needed for simulation	PDF of the hydrogeological parameters, autocorrelation function, and correlation lengths	PDF of hydrogeological parameters, autoregressive parameters, order of the nearest neighbour model to specify a certain auto-correlation function	PDF of hydrogeological parameters, autocorrelation function, and correlation lengths
(2) type of probability density functions (PDF) that can be handled by the method	any probability density function	any probability density function	normal probability density functions (Gaussian fields)

(3) handling auto-correlation structure of any kind	it is straightforward by filling the covariance matrix of the system using the desired auto-correlation function.	a trial and error procedure by adjusting the auto-regressive parameters and the order of the nearest neighbour till it fits the desired auto-correlation; not easy.	it needs a solution of an integral equation that relates the auto-correlation in the field with the auto-correlation on the line process which in some cases is too difficult to solve. In some cases it may be resorted to spectral TBM to tackle this problem.
(4) stationarity of the simulated field	it can handle stationary and non-stationary fields.	it can handle stationary and non-stationary fields, but the method does not guarantee stationarity even if the field is stationary	the field should be stationary

| (5) statistical anisotropy | it handles anisotropy in a simple way by introducing anisotropic auto-correlation functions | it handles a limited range of anisotropy in a simple way by introducing different values of auto-regressive parameters in different directions, but it is difficult to handle highly anisotropic fields. The method does not guarantee isotropy or anisotropy of the simulated field. | it handles anisotropy using transformation method, i.e., an isotropy problem is first transformed into isotropic problem using linear transformation. Second, an isotropic field in the transformed space is generated. Third, transforming back; it needs additional computational efforts and this sort of transform works only for ellipsoidal types of auto-correlation (exponential model). |

(6) matrix operations used	L-U decomposition of the covariance matrix by Cholesky method once, and multiplication of the decomposed matrix by random vector for the generation of each realization.	inversion of a banded matrix (the identity matrix minus spatial lag operator matrix) once, and multiplication by a random vector for each realization.	the line process can be generated by matrix methods, but with a much smaller matrix dimensions or by spectral methods, and generation of a line process several times and its projection onto the problem domain.
(7) storage needed	order of $(\sim N^2)$, where, N is the number of simulated points.	order of $(\sim N^2)$.	order of $(\sim N)$.

(8) computer time and requirements for generating one realization	the major part of the computer time is spent on generating the covariance matrix and its decomposition, once the decomposition is available the generation of one realization is very fast, just by multiplication of the decomposed matrix by uncorrelated random vector.	the major part of the computer time is spent on generating the matrices and its inversion, once the inversion is available the generation of one realization is very fast, just by multiplication of the inverted matrix by uncorrelated random vector.	there is no investment in computer time, but the generation of one realization is "expensive".
(9) ergodicity of the simulated field	non-ergodic because it handles a limited number of simulated points.	non-ergodic because it handles a limited number of simulated points.	ergodic, with lines evenly spaced at pre-specified directions even for a finite number of lines.
(10) conditioning	direct	direct	indirect, by Kriging
(11) efficiency of the method	efficient for a few points and a large number of realizations	efficient for few points and a large number of realizations.	efficient for many points and a few number of realizations.

first order auto-regressive parameters of the *NNG* are chosen to be α_x=.98 and α_y =.5 which generate statistically anisotropic field. Average values of the correlation lengths are calculated from the generated fields over 100 realizations resulting in λ_x=1.2 m and λ_y=0.73 m which show a relatively isotropic field although the ratio of the auto-regressive parameters is about 0.5. These values are used to generate random fields by the other methods (*MVG, TBG*) assuming anisotropic exponential auto-correlations. From Fig.(2.6), the three realizations more or less identical. The auto-correlation function is calculated over ensemble average of 100 realizations and plotted in Fig.(2.7) with the theoretical auto-correlation for comparison which seems reasonably good. Also the calculated *PDF* of a single realization is plotted with the given *PDF* in the same figure which is fairly good as well. The *CPU* time of 100 realization with *NNG* or *MVG* on *PC* with 486 processor was few seconds but for the *TBG* was about 30 minutes.

2.9 Summary and Conclusions

This review highlights some of the progresses achieved in the study of flow and transport in heterogeneous porous media using stochastic approaches. The reason for this approach is two fold. First, the erratic nature of the hydrogeological parameters observed at field data. Second, the uncertainty due to the lack of information about the subsurface structure which is known only at sparse sampled locations. In this chapter the state of the art techniques used for unconditional and conditional stochastic simulation models in porous formations have been reviewed. These techniques are the tools to solve stochastic differential equations of flow and transport in porous media either analytically or numerically.

In the presented techniques, spatial variability of aquifer parameters are modelled as realizations of stationary correlated random fields. These fields are used as input realizations to the classical *PDEs* governing porous flow and transport to evaluate the effect of spatial variability on flow and transport characteristics in porous media. The algorithms used for site characterization and generating heterogeneous descriptions of porous formations are reviewed. These algorithms can be divided broadly into two categories: discrete facies models and continuous models. In discrete facies models, the heterogeneity is described as geometrical shapes of geological units, parameter values to which can later be assigned. In continuous models, the heterogeneity is described as random parameters governed by joint probability distributions.

The techniques of solving *SPDEs* are divided into two main approaches. The analytical approach, which is a sort of linearized solution, such as perturbation method, spectral method, and Green's function method which are used for relatively simple problems and not for practical problems. They are useful to get insight into the problem. The numerical simulation approach (Monte-Carlo), implemented via the application of multivariate method, nearest neighbour method, or turning bands method, are used in more realistic situations where the flow is controlled by complex boundary conditions. It handles the non-linearities of the systems. These approaches are summarized in Fig.(2.3). A brief description of the analytical methods has been given, whereas, an extensive description of some numerical methods has been presented in this chapter. An evaluation of these methods is summarized in a tabular form (see Table (2.1) and Table (2.2)). Advantages, disadvantages and limitations of each of these methods are also presented in these tables. The methods are evaluated on the basis of information available in the literature and by application of these methods to some simple characteristic problems. The following conclusions can be drawn from this review:

(1) For practical applications analytical approaches must be complemented by numerical methods.

(2) Recently developed analytical methods should be evaluated by numerical methods to define their limits of applicability.

(3) Selection of one of these techniques for applications to field problems depends on available field data. Each technique requires specific information. The application depends also on the computer facilities available in terms of storage and speed.

(4) There is no clear proof that one of the numerical techniques is significantly superior to any other, but it is obvious from Table (2.2) that the multi-variate method is more general, although it needs much storage compared with the turning bands method.

(5) Solution of *SPDEs* governing subsurface fluid flow and transport are based on the assumption of Gaussian characteristics and stationarity of the input parameters. Recently, in the hydrogeological field, a considerable attention is devoted to the deviations from these assumptions. It has been shown from geological survey [Krumbein, 1967] that many geological patterns exhibit Markovian properties. Therefore, in this thesis a focus on that area is considered.

(6) The presented stochastic field models are attractive from a statistical point of view, but most of them are less applicable because they are either far from being realistic from a geological point of view or they need intensive hard data (direct measurements of the hydrogeological parameters) to characterize the geological attributes in a proper way. Therefore, this conclusion has motivated the proposed methodology presented in the Chapter 3 which can handle soft geological information to characterize geological features.

(7) Stochastic characterization methods which incorporate the advantages of mosaic facies models and continuous models would be useful to enhance site characterization. This motivated the work described in Chapter 5.

CHAPTER 3

New Methodology for Geological Characterization

3.1 Introduction

Modelling flow and transport in natural formations requires a plausible characterization of the complex heterogeneous structure of these formations. This heterogeneous structure is of great importance to enhance predictions of the hydraulic response of flow and transport such as pressures, seepage velocities, concentration distributions of contaminant plumes and travel times from repositories to accessible environments. The hydraulic response may be used for aquifer management models, remediation strategies of contaminated aquifer systems, and for decision making. At present a variety of techniques are available to describe formation heterogeneity. Some selected ones have been discussed in the previous chapter. Most of these techniques are based on 'hard data' such as direct measurements of flow and transport parameters (hydraulic conductivity, porosity, dispersivity... etc.). Because of economical aspects of engineering projects, there is quite often insufficient hard information about the heterogeneity of geological formations. More indirect qualitative or subjective geological information (soft data) may be available from geological surveys such as geological maps, well logs, bore hole data and geological expertise.

This chapter proposes a practical methodology for modelling geological complexity of natural formations using soft data. The approach presented is an extension of the Markov chain theory used by Krumbein [1967] to synthesize a stratigraphic sequence. Natural aquifers often appear in the form of stratifications due to its sedimentary origin. These stratifications possess long extensions horizontally and relatively small thicknesses vertically. In the present methodology this anisotropic structure is taken into account. The methodology has the following advantages: (1) its theoretical background is simple, (2) it uses soft data that are easy and less expensive to collect, (3) its implementation is simple, (4) conditional simulation to local data is straightforward, (5) it is efficient in terms of computer time and storage, (6) it can be extended to three dimensional problems, (7) it can be used to generate realizations of geological architecture of the subsurface (non-parametric variability) or parametric variability models.

3.2 Background and Description of The Methodology

The sedimentary nature of geological deposits in a fluvial environment exhibits more variability in the vertical direction than in the lateral direction. Stratification or layers are common in natural formations. Due to that fact, a structured anisotropy is manifested, which is reflected in the hydraulic properties of the formation.

Many geological processes display a Markovian property [Krumbein, 1967; Harbaugh and Bonham-Carter, 1970]. The methodology relies on these observations and it assumes that the geological system involves processes that are Markovian, but procedures for statistical testing of the existence of Markovian properties of geological systems have not been developed.

Two coupled chains characterized by two transition probability matrices are used to describe the heterogeneous and the anisotropic structure in natural formations. The horizontal transition between different geological materials is described by a horizontal transition probability matrix. Similarly, the vertical alternation of different geological materials is described by a vertical transition probability matrix. The methodology will adopt the '*Conditional Distribution Approach*', described briefly in Chapter 2 section 2.5.3.2*i* (or refer to Johnson [1987] for more details). The conditional probabilities defined over a grid will be given in terms of transition probabilities of the geological materials that is present in the geological system. In the following sections a brief description of classical and coupled Markov chain theories is presented. It should be mentioned that this methodology is quite different from the other methods described in Chapter 2 (section 2.4.2.1) which use iterative procedures to generate 2D Markovian fields or an implicit formulation like the nearest neighbour model. The methodology is explicit in nature and it uses conditional probabilities.

3.3 Classical Markov Chain Theory

A Markov chain is a probabilistic model that exhibits a special type of dependence: if the state of the system at k-1st observation is given then the state of the system on the kth observation is independent of the preceeding observations earlier than k-1st observation (i.e. if the present is given the future state no longer depends on the past states). In this manner, the state of the system at any time (in case of time series) or space (in case of space series) is stochastically determined by its state at the preceding time or space. Let $Z_0, Z_1, Z_2, \ldots Z_m$ be a sequence of random variables defined on state space $[S_1, S_2, \ldots, S_n]$. The sequence is a Markov chain or Markov process if,

$$Pr[Z_k = S_j \mid Z_{k-1} = S_i, Z_{k-2} = S_n, Z_{k-3} = S_q, ..., Z_0 = S_q] = \\ Pr[Z_k = S_j \mid Z_{k-1} = S_i] = p_{ij}$$ (3.1)

where, the symbol '|' is a symbol for the conditional probability, which means the probability of occurrence of a certain state given that other states have already occurred and p_{ij} is the transition probability from state S_i to state S_j.

The Markov chain model can be classified as a sequence-based model mentioned in Chapter 2. It can be applied in geology to model discrete variables such as a lithology or a facies. The model does not use variograms or autocovariance functions to quantify the spatial structures as most of the available models do (see Chapter 2). Instead it uses conditional probabilities. Conditional probabilities have the advantage that they are interpretative geologically much easier than variogram or autocovariance functions. This is the reason for their popularity in the geological community [Davis, 1973]. There are also limitation and sparsity of the available field data to infer reliable variograms or auto-covariances.

Markov chain as presented in this section is known as the simple or first-order form of a chain. Furthermore, the probabilities associated with the transitions between the states are stationary or homogeneous (i.e. do not change with translation). Generalizations are possible where the conditional dependence involves states earlier than the preceeding state (longer memory). This generalization is known as a complex or higher-order chain. Another generatization is the nonstationary or nonhomogeneous chains where the transition probabilities change with time or space. These kinds of generalizations need more data than the first-order homogeneous ones. In this chapter only first-order stationary chain is considered.

3.3.1 Markov Transition Probability Matrix

In one-dimensional problems, a Markov chain is described by a single transition probability matrix [Krumbein, 1967]. Transition probabilities are defined as a relative frequency of transitions from a certain state to another state in a system that consists of a number of states. These transitions can be arranged in a square matrix form,

$$p = \begin{bmatrix} p_{11} & p_{12} & \cdots & p_{1n} \\ p_{21} & p_{22} & \cdots & p_{2n} \\ & & \cdot & \\ p_{n1} & p_{n2} & \cdots & p_{nn} \end{bmatrix}$$ (3.2)

where, p_{kl} signifies the probability of transition from state S_k to state S_l, and n is the number of states in the system.

Thus the probability of a transition from S_1 to S_1, S_2, ..., S_n is given by p_{1l}, $l = 1,2,.....n$ in the first row and so on. The matrix p has to fulfil specific properties: (1) its elements can not be negative, $p_{kl} > 0$; (2) the elements of each row sum up to one or,

$$\sum_{l=1}^{n} p_{kl} = 1 \qquad (3.3)$$

3.3.2 Marginal (Fixed) Probabilities

The transition probabilities considered in the previous section are called single step transition, which means, the transition from a state to another takes place in a single step. If one considers m-steps of transitions which means a transition form a state to another takes place in m-steps. The m-steps transition probabilities can be obtained by multiplying the single step transition probability matrix by itself m times. The successive multiplications lead to identical rows. In other words, at some point in time or space the probability of occurrence of a state is independent of the initial state in the sequence. The probability of occurence w_l, where, $l = 1,2,...n$ such that,

$$\lim_{m \to \infty} p_{kl}^{(m)} = w_l \qquad (3.4)$$

is called marginal or fixed probabilities. These are the probabilities of being in a certain state say S_l, regardless the state of the previous observation. The w_l are no longer dependent of the initial state S_k. The limiting probabilities may also be found by solving the equations,

$$\sum_{k=1}^{n} w_k \, p_{kl} = w_l \, , \qquad l = 1 \, , \, n \qquad (3.5)$$

subjected to the conditions,

$$w_l \geq 0 \quad , \textit{for all } l$$

$$\sum_{l=1}^{n} w_l = 1 \qquad (3.6)$$

this leads to a so-called marginal distribution, which can be written in a vector form as,

$$w = \begin{bmatrix} w_1 \\ w_2 \\ . \\ w_n \end{bmatrix} \qquad (3.7)$$

where, w_l is the marginal probability of state l, $l = 1,2,3...n$

This property of the chain is a manifestation of 'memory', which the state of the system at certain time or space remembers the preceeding state, but the state's influence diminishes with time or distance until it is forgotten.

3.4 Coupled Markov Chain Theory

The coupled chain describes the joint behaviour of pairs of independent systems, each evolving according to the laws of the classical Markov chain [Billingsley, 1995]. Consider two Markov chains X_m, Y_m, $m = 0,1,2,....$, both defined on the state space $[S_1, S_2, ..., S_n]$ and having the positive transition probability defined as,

$$Pr[X_{m+1} = S_l , Y_{m+1} = S_k \mid X_m = S_i , Y_m = S_j] = p(ij,lk) \qquad (3.8)$$

The coupled transition probability p(ij,lk) is given by,

$$p(ij,lk) = p_{il} \cdot p_{jk} \qquad (3.9)$$

These transition probabilities form a stochastic matrix. The marginal distribution of the coupled chain is given by,

$$w(lk) = w_l \cdot w_k \qquad (3.10)$$

3.5 Joint Probability on Lattice

Extensions of Markov chain theory to higher dimensions can be obtained through the application of coupled chain theory given in (section 3.5). In this section a generalization to two dimensions is considered.

Consider a two dimensional domain of cells as shown in Fig.(3.1). Each cell has a row number j and a column number i. Consider also a given number of geological materials say n, such as sand, clay, peat, etc. These geological materials

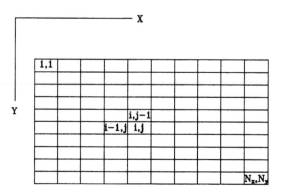

Fig.(3.1) Cell Numbering in Two Dimensional Domain.

are coded, for example, 1 for sand, 2 for clay and 3 for peat, etc.. This coding is translated later to hydraulic conductivity values corresponding to each geological material. The word '*state*' as used in this text, now describes a certain unit, lithology, or geological material in the geological system. The joint probability of the lattice process can be expressed mathematically as,

$$Pr \ (z_{1,1} = S_1, \ z_{2,1} = S_2,..., \ z_{i,j} = S_k, \ z_{i-1,j} = S_l, \ z_{i,j-1} = S_q..., \ z_{N_x,N_y} = S_p)$$

where, S_k is a state of cell (i,j), which is one of the n states describing the geological system, N_x is the maximum number of cells in the horizontal direction, N_y is the maximum number of cells in the vertical direction.

Expansion of the joint probability function using the axiom of conditional probability function 'Conditional Distribution Approach' can be done. One of the many possible expansions is,

$$Pr \ (z_{1,1} = S_1, \ z_{2,1} = S_2,..., \ z_{i,j} = S_k, \ z_{i-1,j} = S_q, \ z_{i,j-1} = S_l...$$
$$, \ z_{N_x,N_y} = S_p) =$$
$$Pr \ (z_{N_x,N_y} = S_p| \ z_{N_x-1,N_y} = S_q, \ z_{N_x,N_y-1} = S_l,..., \ z_{1,1} = S_1) \ ...$$
$$Pr \ (z_{i,j} = S_k| \ z_{i-1,j} = S_l, \ z_{i,j-1} = S_r, \ ..., \ z_{1,1} = S_1) \ ...$$
$$Pr \ (z_{2,2} = S_f| \ z_{1,2} = S_3, \ z_{2,1} = S_2, \ z_{1,1} = S_1) \ .$$
$$Pr \ (z_{2,1} = S_2| \ z_{1,1} = S_1) \ . \ Pr \ (z_{1,1} = S_1)$$

(3.11)

where, $Pr(z_{1,1} = S_1)$ is the marginal probability of state S_1

One can get a more simple form of Eq.(3.11) by introducing some nearest neighbour property according to Markov chain theory [Brook, 1964], and by making use of two first order Markov chains for the geological variability in horizontal and vertical directions. Eq.(3.11) becomes,

$$Pr\ (z_{1,1} = S_1,\ z_{2,1} = S_2,...,\ z_{i,j} = S_k,\ z_{i-1,j} = S_l,$$
$$z_{i,j-1} = S_q...,\ z_{N_x,N_y} = S_p) =$$
$$Pr\ (z_{N_x,N_y} = S_p\ |\ z_{N_x-1,N_y} = S_l,\ z_{N_x,N_y-1} = S_a)...$$
$$Pr\ (z_{i,j} = S_k\ |\ z_{i-1,j} = S_l,\ z_{i,j-1} = S_f)...$$
$$Pr\ (z_{2,2} = S_4\ |\ z_{2,1} = S_3,\ z_{1,2} = S_2)\ .$$
$$Pr\ (z_{N_x,1} = S_d\ |\ z_{N_x-1,1} = S_g)...$$
$$Pr\ (z_{2,1} = S_2\ |\ z_{1,1} = S_1)\ .$$
$$Pr\ (z_{1,N_y} = S_r\ |\ z_{1,\ N_y-1} = S_x)\ ...$$
$$Pr\ (z_{1,2} = S_2\ |\ z_{1,1} = S_1)\ .\ Pr\ (z_{1,1} = S_1)$$

$$(3.12)$$

how this formula is used to generate realizations of geological structures possessing a Markovian property is explained in (section 3.7).

3.5.1 Classical First Order Markov Chain on Domain Boundaries

In Eq.(3.12), the conditional probabilities used to generate the boundaries are described by classical first order Markov chain. A conditional probability in the upper horizontal boundary is,

$$Pr\ (z_{i,1} = S_k\ |\ z_{i-1,1} = S_l),\qquad i = 2,3,4......N_x$$

Similarly, a conditional probability in left vertical boundary is,

$$Pr\ (z_{1,j} = S_m\ |\ z_{1,j-1} = S_q),\qquad j = 2,3,4,......N_y$$

Conditional probabilities on these boundaries are given in terms of Markov transition probabilities for horizontal and vertical directions respectively. If transition probability matrices for two chains are notated as,

$$p^d = \begin{bmatrix} p_{11}^d & p_{12}^d & .... & p_{1n}^d \\ p_{21}^d & p_{22}^d & .... & p_{2n}^d \\ & & \cdot & \\ p_{n1}^d & p_{n2}^d & .... & p_{nn}^d \end{bmatrix} \qquad (3.13)$$

where, d is an indicator for the direction $d = v$ for vertical direction and $d = h$ for horizontal direction. This is expressed as,

71

$$Pr\ (z_{i,1} = S_k \mid z_{i-1,1} = S_l) = p_{lk}^{h}$$
$$Pr\ (z_{1,j} = S_m \mid z_{1,j-1} = S_q) = p_{qm}^{v}$$

(3.14)

3.5.2 Coupled Markov Chain on Lattice

The conditional probabilities for the rest of the cells on the lattice are calculated by coupling both horizontal and vertical chains. A general formula of the conditional probability at any arbitrary cell on the lattice (see Fig.(3.1)) is

$$Pr\ (z_{i,j} = S_k \mid z_{i-1,j} = S_l,\ z_{i,j-1} = S_m)$$

where, $i = 2,3,..........N_x$, and $j = 2,3,..........N_y$. This term can be expanded into two terms based on the probability axiom assuming independency of the horizontal and vertical chains and choosing the set of probabilities where the same state can be obtained from both chains. This is expressed as,

$$Pr\ (z_{i,j} = S_k \mid z_{i-1,j} = S_l,\ z_{i,j-1} = S_m) =$$
$$Pr\ (z_{i,j} = S_k \mid z_{i-1,j} = S_l)\ .\ Pr\ (z_{i,j} = S_k \mid z_{i,j-1} = S_m)$$

(3.15)

which is factorized again into first-order terms, one is conditioned horizontally and the other is conditioned vertically. The value of this conditional probability is calculated by multiplying the corresponding terms from the horizontal and vertical transition probability matrices (see Eq.(3.9)). One has to select only the cases where l and k in Eq.(3.9) have the same outcome.

$$p(lm, kk) = p_{lk}^{h} \cdot p_{mk}^{v}$$

(3.16)

Since these cases constitute only set of all possible outcomes then their probabilities do not sum up to one. Therefore, during Monte-Carlo sampling to realize the coupled chain the transitions probabilities should be normalized. The normalization procedure is explained in (see section 3.7).

3.6 Inference of Statistical Properties from a Geological System

A Markov chain is completely described when the state space, transition probabilities and initial probabilities are given. For a geological system represented by a Markov chain this means that firstly, the set of possible states of the system $[S_1, S_2, ...,$ $S_n]$ must be determined or defined. Secondly, the probability, p_{kl}, of going from a state S_l to state S_k in one interval must be determined or estimated. Finally, the marginal probabilities w_l are determined either by estimation or calculation. In practical applications, transition and marginal probabilities of a geological system can be estimated from well logs, boreholes, surface and subsurface mapping or from geological image synthesized by information derived from geologically similar sites or analogous outcrops. Some applications are considered in the following sections.

3.6.1 Parameter Estimation from Well Logs and Geological Maps

The core of stochastic modelling in geology is to get a plausible description of the subsurface with limited amount of data. A detailed description of deriving statistical properties from well logs and geological maps is illustrated. Fig.(3.2) shows the steps followed to generate realizations of geological patterns by this model where the parameters needed are estimated from well logs data and geological maps. Parameter estimation procedures are explained below.

3.6.1.1 Estimation of Vertical Transition Probabilities

The vertical transition probability matrix can be estimated from well logs. The tally matrix of vertical transitions is obtained by superimposing a vertical line with equidistant points along the well with a chosen sampling interval. The transition frequencies between the states are calculated by counting how many times a given state say S_k is followed by itself or the other states say S_l in the system and then dividing by the total number of transitions,

$$p_{kl}^v = \frac{T_{kl}^v}{\sum_{l=1}^{n} T_{kl}^v} \tag{3.17}$$

where, T_{kl}^v is the number of observed transitions from state S_k to state S_l in the vertical direction. The cumulative vertical transition probability matrix (which is defined as, given that the neighbour state has state S_k, P_{kl}^v gives the conditional probability that the cell itself is either $S_1, S_2, S_3, \ldots S_l$.) is computed by the formula,

$$P_{kl}^v = \sum_{m=1}^{l} p_{km}^v \tag{3.18}$$

where, P_{kl}^v is the cumulative vertical transition probability distribution from state S_k to state S_l and n is the number of states in the system. In this cumulative form the probability values in each row progressively sum to 1.0. This cumulative distribution is used to realize the chain (see section 3.7).

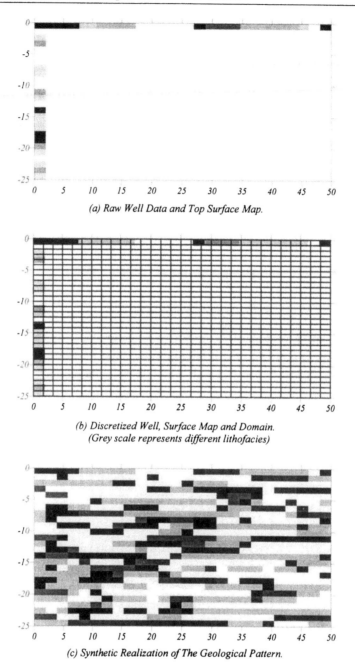

(a) Raw Well Data and Top Surface Map.

(b) Discretized Well, Surface Map and Domain.
(Grey scale represents different lithofacies)

(c) Synthetic Realization of The Geological Pattern.

Fig.(3.2) Parameter Estimation and Simulation from Well Data and Geological Maps.

74

3.6.1.2 Estimation of Horizontal Transition Probabilities

The horizontal transition probability matrix can be estimated from geological maps describing the formation extensions in horizontal plane. These maps may be obtained from geological surveying. On the map plan, a transect is defined where the subsurface profile is required. On that transect, a similar procedure is performed as in the vertical transitions. A horizontal line with equidistance points at a proper sampling interval for the horizontal direction is superimposed. The transitions between different states are counted and the horizontal transition probability matrix is computed in a similar way as for the vertical transitions. Each element in the horizontal transition matrix is calculated by a similar formula as Eq.(3.17) with superscript h instead of v,

$$p_{kl}^{h} = \frac{T_{kl}^{h}}{\sum_{l=1}^{n} T_{kl}^{h}} \qquad (3.19)$$

where, T_{kl}^{h} is the number of observed transitions from state S_k to state S_l in the horizontal direction. The cumulative horizontal transition probability distribution matrix is computed by adding each probability value to each succeeding value, moving from left to right within each row. This is expressed mathematically as,

$$P_{kl}^{h} = \sum_{m=1}^{l} p_{km}^{h} \qquad (3.20)$$

where, P_{kl}^{h} is the cumulative horizontal transition probability distribution from state S_k to state S_l. This matrix has the same property as in Eq.(3.18) that the probability values in each row progressively sum to 1.

3.6.1.3 Estimation of Marginal Probabilities

The marginal probability is defined as the probability of occurrence of a given state in the system regardless the states of neighbouring cells. One can get the marginal probabilities, w_l, by summing the columns of the transition frequency matrix and dividing these column totals by the grand total of the transitions as [Davis, 1973],

$$w_l = \frac{\sum\limits_{k=1}^{n} T_{kl}}{\sum\limits_{k=1}^{n} \sum\limits_{l=1}^{n} T_{kl}} \tag{3.21}$$

In case of using two transition probability matrices in the horizontal and vertical directions Eq.(3.21) can be obtained by,

$$w_l = \frac{\sum\limits_{k=1}^{n} T_{kl}^{h} + \sum\limits_{k=1}^{n} T_{kl}^{v}}{\sum\limits_{k=1}^{n} \sum\limits_{l=1}^{n} T_{kl}^{h} + \sum\limits_{k=1}^{n} \sum\limits_{l=1}^{n} T_{kl}^{v}} \tag{3.22}$$

where, T_{kl}^{h} is the horizontal transition frequency between state S_k and state S_l, and T_{kl}^{v} is the vertical transition frequency between state S_k and state S_l.

The cumulative marginal probability distribution vector of the different states can also be computed by,

$$W_l = \sum\limits_{k=1}^{l} w_k \tag{3.23}$$

The marginal probability distribution vector has the property that its components progressively sum to one.

3.6.1.4 Estimation of Sampling Intervals

Choice of the optimal sampling intervals to estimate the transitions is not an easy task. Sampling intervals less than proper ones will yield a large number of cells and high resolution, whereas large sampling intervals will yield less resolution. There is a tradeoff between highly resoluted image with fine grid and the additional counting effort and computer storage requirements that it entails versus the low resoluted image of course grid with its accompanying reduction in counting effort and computer storage requirements. Perhaps the proper sampling interval in the vertical direction would be less than or equal to the minimum thickness of the geologic unit found in the well log to be reasonably reproduced. Similarly, the proper sampling interval in the horizontal direction would be less than or equal to the minimum length of a geological unit found in the map.

3.6.2 Parameter Estimation from Geological Expertise

In this application, the procedure may be as follows. Information about the transition probabilities may be provided directly from geological experience, geologically similar formations but better known or analogous outcrops. A subjectively hand-drawn cross-section can be made. A two-dimensional grid with a proper sampling intervals, as discussed earlier, in the horizontal and vertical directions is superimposed over the geological image. Fig.(3.3) shows the steps followed for this simulation. Transition frequencies between states of the system can be obtained by counting transitions along successive grid lines in both directions of the grid. The transition frequencies are counted separately along both horizontal and vertical directions. Frequencies thus obtained can be transformed into estimates of transition probabilities discussed in the previous section. These probabilities are introduced to the developed simulation program to produce number of realizations of the geological pattern. These realizations can be used in Monte-Carlo studies (see Chapter 2 section 2.5.2) for risk analysis. This will be explained in more details in Chapter 7.

3.7 Implementation

A procedure for Monte-Carlo sampling to invoke this methodology is presented. Fig.(3.1) will be considered during description of the algorithm. A procedure for unconditional simulation involves the following set of instructions using Eq.(3.12) in a sequential order:

Step 1: The two dimensional domain is discretized using proper sampling intervals see section (4).

Step 2: Select an initial state at cell (1,1), the upper left corner, from the marginal distribution of the states, w_l, $l = 1,2,...n$ if it is known, or randomly from the given states in the system, if the distribution is not known. The procedure is simple if one wishes to choose between discrete states. The choice between the different states can be realized by solving the following equation,

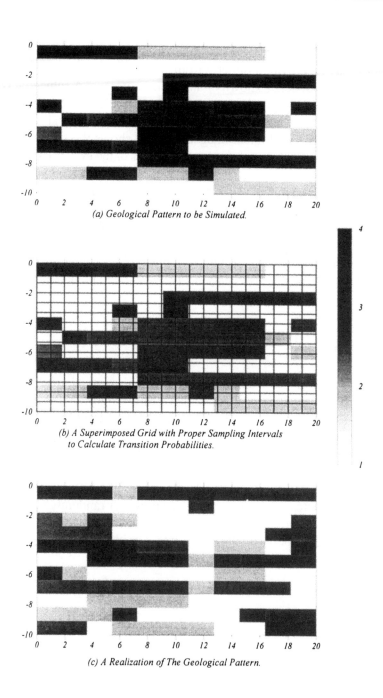

(a) Geological Pattern to be Simulated.

(b) A Superimposed Grid with Proper Sampling Intervals
to Calculate Transition Probabilities.

(c) A Realization of The Geological Pattern.

Fig.(3.3) Parameter Estimation and Simulation from Geological Expertise.

$$\sum_{l=1}^{k-1} w_l < r_1 \le \sum_{l=1}^{k} w_l \ , \qquad\qquad k = 2,....n \qquad (3.24)$$

where, r_1 is a uniform random number between 0 and 1.

Step 3: Start from cell $(1,1)$, generate horizontal boundary cells $(i,1)$, $i = 2,....$, N_x from the conditional distribution $f(z_{i,1} = S_k | z_{i-1,1} = S_l)$, given that the state at $(i-1,1)$ is known. Select the succeeding state at $(i,1)$ at random from the row corresponding to S_l in the cumulative horizontal transition probability distribution matrix. A realization of the succeeding state S_k is obtained by solving,

$$\sum_{q=1}^{k-1} P_{lq}^{h} < r_2 \le \sum_{q=1}^{k} P_{lq}^{h} \ , \qquad k = 2,...n \ , \ l = 1,...n \qquad (3.25)$$

where, r_2 is a uniform random number between 0 and 1.

Step 4: Similarly, from cell $(1,1)$, generate vertical boundary cells $(1,j)$, $j = 2,....$, N_y from the conditional distribution $f(z_{1,j} = S_k | z_{1,j-1} = S_m)$, given that the state at $(1,j-1)$ is known. Select the succeeding state at $(1,j)$ at random from the row corresponding to S_m in the cumulative vertical transition probability distribution matrix. A realization of the succeeding state S_k is obtained by solving a similar equation to Eq.(3.25) but for the vertical chain.

Step 5: Generate the rest of the cells with numbers (i,j), $i = 2,....$, N_x, and $j = 2,....$, N_y in the domain rowwise using information from the horizontal and vertical directions. This can be performed using the conditional distribution $f(z_{i,j} = S_k | z_{i-1,j} = S_l, z_{i,j-1} = S_m)$, given that the state at $(i-1,j)$ and $(i,j-1)$ are known, which is calculated by Eq.(3.16), simply by multiplying the corresponding terms from the vertical and horizontal transition probability matrices. The resulting product produces a three indices transition probability p_{ijk}. At each state S_i from the horizontal neighbouring cell and S_j from the vertical neighbouring cell one can determine the succeeding state S_k. The probability of choosing $S_1, S_2, ...$ or S_m for S_k can be realized by solving the following equation for S_k,

$$\frac{\sum_{l=1}^{k-1} P_{ijl}}{\sum_{l=1}^{n} P_{ijl}} < r \le \frac{\sum_{l=1}^{k} P_{ijl}}{\sum_{l=1}^{n} P_{ijl}} \ , \ k = 2,...n, \ i \ , j = 1,...n \qquad (3.26)$$

Step 6: The procedure stops after having visited all cells in the domain.

3.8 Conditional Simulation on Field Data

Conditional simulation at local measurements is straightforward. Instead of generating upper horizontal boundary and left vertical boundary from the derived transition probability matrices, one can assign at each cell on the boundaries the corresponding state from the well log or bore hole data and given geological map. The rest of the domain cells are generated using step 5 described in the algorithm (section 3.7). Thus the produced image is conditioned to the field data. Conditioning on more than one well needs further investigation which is not covered in this research.

3.9 Simulation of Inclined Bedding

The same methodology can be adjusted to address inclined bedding in the formation structure or for inclined flow. A slight modification should be applied to the probabilistic operator, Eq.(3.15). The following conditional probability scheme could be adopted,

$$Pr\ (z_{i,j+1} = S_k \mid z_{i-1,j} = S_l\ ,\ z_{i+1,j} = S_m) =$$
$$Pr\ (z_{i,j+1} = S_k \mid z_{i-1,j} = S_l).\ Pr\ (z_{i,j+1} = S_k \mid z_{i+1,j} = S_m) \qquad (3.27)$$

where, $i = 2,3,..........N_x$, and $j = 2,3,..........N_y$. The transition probabilities that will be used to determine the conditional probabilities should be calculated in the direction of the bedding and normal to it. The bedding angle is related to the sampling intervals by,

$$\tan \theta = \frac{\delta y}{\delta x} \qquad (3.28)$$

where, θ is the bedding angle, δx and δy are the sampling interval in x and y direction respectively.

3.10 Testing The Methodology

Some numerical tests have been performed to check the results of this methodology. Extreme cases are considered: uncorrelated and perfectly correlated fields.

3.10.1 Test (i): Independent Markovain Random Field (Uncorrelated Field)

Fig.(3.4a) shows the case of uncorrelated field in both horizontal and vertical directions. This has been done by putting the diagonal elements of the transition probability matrix for both horizontal and vertical directions equal to zero (see Table 3.1). This means that, there is no possibility for a transition from one state to itself. It appears in the figure that each state is always surrounded by different states from the four neighbours. The given and the simulated statistics displayed in Table (3.1) shows very good agreement. Fig.(3.4b) shows the same results for a field generated with the same transitions but with inclination to the horizontal with 45°. The statistics of this case are given in Table (3.2).

3.10.2 Test (ii): Perfect Horizontal Stratification (Perfectly Correlated Field in Horizontal)

Fig.(3.4c) shows another test. This case is for checking the other extreme of the previous test where the field is perfectly correlated in the horizontal direction. This can be done by using a unit matrix for the horizontal transition probability (i.e. the diagonal elements are one which means that the probability for a transition from a state to itself horizontally is 100% and there no possibility for a transition to other state) given in Table (3.3). The statistical results show perfect agreement for the horizontal transitions but there are small discrepancies in the vertical due to the fact that there is no sufficient number of transitions in the vertical direction to reproduce the vertical transition probability matrix very well. Fig.(3.4c) displays a geological stratigraphic sequence similar to the one synthesised by Krumbein [1967] using a single transition probability matrix.

3.10.3 Test (iii): Perfect Vertical Strips (Perfectly Correlated Field in Vertical)

Fig.(3.4d) shows the last test. This test is similar to the previous one. It is performed to check the perfect correlation in the vertical direction. In this case a unit matrix for the vertical transition probability is used (i.e. the diagonal elements are one which means that the probability for a transition from a state to itself vertically is 100% and there no possibility for a transition to other state). As it is clear from Table (3.4), the vertical transition probability matrix is reproduced exactly. For the horizontal transitions there are minor fluctuations due to the small number of transitions.

Table (3.1) Statistics of Fig.(3.4a).

Sampling interval in X-axis (m)=   1.0
Sampling interval in Y-axis (m)=   1.0

Input Statistics   Calculated Statistics

Horizontal Transition Probability Matrix

State	1	2	3	4	State	1	2	3	4
1	0.000	0.340	0.330	0.330	1	0.000	0.360	0.322	0.318
2	0.340	0.000	0.330	0.330	2	0.332	0.000	0.332	0.336
3	0.330	0.340	0.000	0.330	3	0.348	0.326	0.000	0.326
4	0.340	0.330	0.330	0.000	4	0.330	0.338	0.332	0.000

Vertical Transition Probability Matrix

State	1	2	3	4	State	1	2	3	4
1	0.000	0.340	0.330	0.330	1	0.000	0.384	0.310	0.305
2	0.340	0.000	0.330	0.330	2	0.319	0.000	0.343	0.339
3	0.330	0.340	0.000	0.330	3	0.353	0.312	0.000	0.336
4	0.340	0.330	0.330	0.000	4	0.341	0.324	0.336	0.000

Table (3.2) Statistics of Fig.(3.4b).

Sampling interval in X-axis (m)=   1.0
Sampling interval in Y-axis (m)=   1.0

Input Statistics   Calculated Statistics

XX' - XX' Transition Probability Matrix

State	1	2	3	4	State	1	2	3	4
1	0.000	0.340	0.330	0.330	1	0.000	0.347	0.353	0.300
2	0.340	0.000	0.330	0.330	2	0.347	0.000	0.323	0.330
3	0.330	0.340	0.000	0.330	3	0.340	0.347	0.000	0.313
4	0.330	0.340	0.330	0.000	4	0.303	0.356	0.341	0.000

YY' - YY' Transition Probability Matrix

State	1	2	3	4	State	1	2	3	4
1	0.000	0.340	0.330	0.330	1	0.002	0.354	0.340	0.304
2	0.340	0.000	0.330	0.330	2	0.342	0.002	0.338	0.319
3	0.330	0.340	0.000	0.330	3	0.332	0.349	0.002	0.318
4	0.330	0.340	0.330	0.000	4	0.318	0.345	0.335	0.003

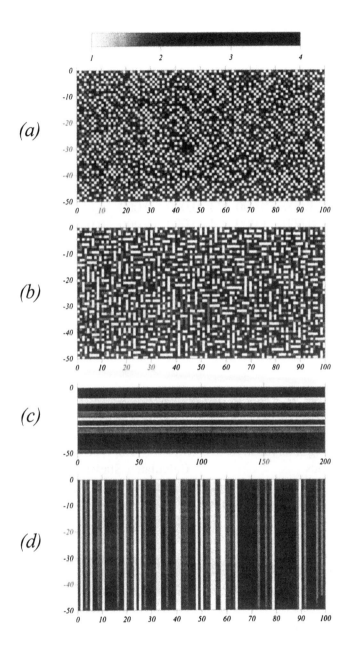

Fig.(3.4) Tests: (a) Uncorrelated Field in Horz. and Vert. Directions. (b) Uncorrelated Field with 45°. (c) Perfect Correlation in Horz. (d) Perfect Correlation in Vert..

Table (3.3) Statistics of Fig.(3.4c).

Sampling interval in X-axis (m)=   1.0
Sampling interval in Y-axis (m)=   1.0

Input Statistics   Calculated Statistics

Horizontal Transition Probability Matrix

State	1	2	3	4	State	1	2	3	4
1	1.000	0.000	0.000	0.000	1	1.000	0.000	0.000	0.000
2	0.000	1.000	0.000	0.000	2	0.000	1.000	0.000	0.000
3	0.000	0.000	1.000	0.000	3	0.000	0.000	1.000	0.000
4	0.000	0.000	0.000	1.000	4	0.000	0.000	0.000	1.000

Vertical Transition Probability Matrix

State	1	2	3	4	State	1	2	3	4
1	0.600	0.100	0.200	0.100	1	0.571	0.000	0.143	0.286
2	0.100	0.700	0.100	0.100	2	0.091	0.818	0.000	0.091
3	0.150	0.150	0.400	0.300	3	0.111	0.111	0.222	0.556
4	0.100	0.200	0.200	0.500	4	0.043	0.087	0.261	0.609

Table (3.4) Statistics of Fig.(3.4d).

Sampling interval in X-axis (m)=   1.0
Sampling interval in Y-axis (m)=   1.0

Input Statistics   Calculated Statistics

Horizontal Transition Probability Matrix

State	1	2	3	4	State	1	2	3	4
1	0.250	0.250	0.250	0.250	1	0.211	0.211	0.211	0.368
2	0.250	0.250	0.250	0.250	2	0.280	0.240	0.240	0.240
3	0.250	0.250	0.250	0.250	3	0.179	0.143	0.286	0.393
4	0.250	0.250	0.250	0.250	4	0.107	0.393	0.357	0.143

Vertical Transition Probability Matrix

State	1	2	3	4	State	1	2	3	4
1	1.000	0.000	0.000	0.000	1	1.000	0.000	0.000	0.000
2	0.000	1.000	0.000	0.000	2	0.000	1.000	0.000	0.000
3	0.000	0.000	1.000	0.000	3	0.000	0.000	1.000	0.000
4	0.000	0.000	0.000	1.000	4	0.000	0.000	0.000	1.000

3.11 Some Geological Simulations

In this section some samples of natural geological formations that could be expected in the field will be generated using the present technique. Fig.(3.5) to Fig.(3.7) display some of the different types of geological formations which can be simulated using the present model. All applications illustrated below involve comparison of the actual statistics that used as an input to the model with statistics derived from the generated images. The statistics of the figures are tabulated from Table (3.5) to Table (3.13).

Fig.(3.5a) shows two-phase geological materials, such as, sand-shale formation frequently used in Petroleum reservoirs. Fig.(3.5b) displays a four states stratified geological system with quite small thicknesses which could be deposited in a sedimentary basin. Fig.(3.5c) shows a four states stratified geological pattern with relatively variable thicknesses of the states present in a system. Fig.(3.6a) shows a four states geological formation with a single state possessing large thickness and long extension in comparison to the rest. Fig.(3.6b) shows a four states geological system of large scale formation structure with units possessing relatively big dimensions. Fig.(3.6c) shows a four states geological system of base rock with some other sedimentary layers deposited over it. Fig.(3.7a), Fig.(3.7b) and Fig.(3.7c) show a four states geological system of cross-bedding with different inclinations with respect to the horizontal axis e.g. -45, -7., and 26.5 degrees respectively.

3.12 Parametric Variability Models

The proposed stochastic methodology can address also parameter variability. It can be done by defining for each individual state in the model a prescribed class of the parameter under study. For instance, generation of realizations of hydrogeological parameter fields can be done by assigning to each state in the system a unique value of a hydrogeological parameter. Another way, is to assign to each state a specified class of the parameter under study, i.e., each state represents a certain range of the hydrogeological parameter, for example, a permeability in the range of 1 to 5 m/day for the first class and 5 to 10 m/day for the second class and so on. The generated realizations of the parameters are characterized by conditional probabilities rather than auto-covariances.

Table (3.5) Statistics of Fig.(3.5a).

Sampling interval in X-axis (m)= 10.0
Sampling interval in Y-axis (m)= 5.0

Input Statistics		Calculated Statistics	
Horizontal Transition Probability Matrix			
0.980	0.020	0.983	0.017
0.020	0.980	0.023	0.977
Vertical Transition Probability Matrix			
0.500	0.500	0.567	0.433
0.500	0.500	0.581	0.419

Table (3.6) Statistics of Fig.(3.5b).

Sampling interval in X-axis (m)= 10.0
Sampling interval in Y-axis (m)= 5.0

Input Statistics				Calculated Statistics			
Horizontal Transition Probability Matrix							
0.970	0.010	0.010	0.010	0.970	0.012	0.010	0.009
0.005	0.980	0.005	0.010	0.005	0.984	0.004	0.007
0.020	0.020	0.940	0.020	0.017	0.016	0.951	0.016
0.010	0.010	0.010	0.970	0.010	0.013	0.012	0.965
Vertical Transition Probability Matrix							
0.250	0.250	0.250	0.250	0.307	0.387	0.138	0.168
0.250	0.250	0.250	0.250	0.188	0.431	0.146	0.235
0.250	0.250	0.250	0.250	0.253	0.384	0.214	0.148
0.250	0.250	0.250	0.250	0.290	0.336	0.139	0.235

Table (3.7) Statistics of Fig.(3.5c).

Sampling interval in X-axis (m)= 10.0
Sampling interval in Y-axis (m)= 5.0

Input Statistics				Calculated Statistics			
Horizontal Transition Probability Matrix							
0.950	0.005	0.005	0.040	0.956	0.004	0.001	0.039
0.010	0.970	0.010	0.010	0.016	0.966	0.005	0.013
0.020	0.010	0.960	0.010	0.058	0.032	0.894	0.016
0.010	0.010	0.010	0.970	0.012	0.008	0.004	0.976
Vertical Transition Probability Matrix							
0.700	0.150	0.100	0.050	0.746	0.130	0.038	0.086
0.200	0.500	0.100	0.200	0.154	0.458	0.025	0.363
0.300	0.200	0.300	0.200	0.236	0.202	0.081	0.480
0.100	0.200	0.100	0.600	0.068	0.138	0.030	0.765

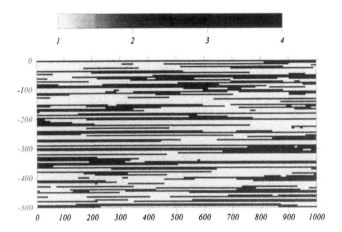

(a) Geological Pattern with Two Phase Materials.

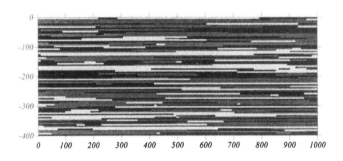

(b) Relatively Thin Stratified Formation.

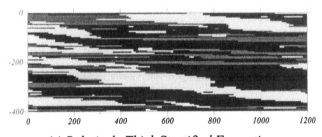

(c) Relatively Thick Stratified Formation.

Fig.(3.5) Some Geological Patterns: (a) Two Phase Geological Materials.
(b) Realtively Thin Stratifications. (c) Relatively Thick Layers.

87

Table (3.8) Statistics of Fig.(3.6a).

Sampling interval in X-axis (m)=			10.0				
Sampling interval in Y-axis (m)=			10.0				
Input Statistics				Calculated Statistics			
Horizontal Transition Probability Matrix							
0.900	0.050	0.040	0.010	0.959	0.023	0.016	0.002
0.050	0.800	0.100	0.050	0.057	0.834	0.081	0.028
0.020	0.020	0.940	0.020	0.032	0.030	0.919	0.019
0.010	0.010	0.010	0.970	0.021	0.010	0.007	0.962
Vertical Transition Probability Matrix							
0.800	0.100	0.050	0.050	0.896	0.026	0.030	0.048
0.100	0.600	0.200	0.100	0.162	0.496	0.253	0.089
0.100	0.300	0.500	0.100	0.102	0.215	0.540	0.143
0.100	0.100	0.200	0.600	0.067	0.044	0.237	0.652

Table (3.9) Statistics of Fig.(3.6b).

Sampling interval in X-axis (m)=			10.0				
Sampling interval in Y-axis (m)=			5.0				
Input Statistics				Calculated Statistics			
Horizontal Transition Probability Matrix							
0.900	0.030	0.030	0.040	0.924	0.017	0.047	0.013
0.010	0.970	0.010	0.010	0.022	0.960	0.016	0.002
0.010	0.020	0.960	0.010	0.024	0.010	0.963	0.003
0.040	0.040	0.010	0.910	0.093	0.053	0.099	0.755
Vertical Transition Probability Matrix							
0.970	0.010	0.010	0.010	0.939	0.020	0.038	0.004
0.040	0.900	0.030	0.030	0.022	0.940	0.033	0.006
0.020	0.010	0.960	0.010	0.017	0.007	0.972	0.005
0.040	0.040	0.010	0.910	0.134	0.043	0.062	0.761

Table (3.10) Statistics of Fig.(3.6c).

Sampling interval in X-axis (m)=			10.0				
Sampling interval in Y-axis (m)=			5.0				
Input Statistics				Calculated Statistics			
Horizontal Transition Probability Matrix							
0.400	0.200	0.300	0.100	0.924	0.030	0.041	0.006
0.010	0.970	0.010	0.010	0.076	0.914	0.006	0.005
0.010	0.020	0.960	0.010	0.093	0.026	0.876	0.005
0.010	0.005	0.005	0.980	0.067	0.013	0.007	0.913
Vertical Transition Probability Matrix							
0.990	0.005	0.001	0.004	0.981	0.014	0.004	0.002
0.200	0.400	0.300	0.100	0.039	0.555	0.343	0.063
0.200	0.300	0.400	0.100	0.053	0.367	0.526	0.054
0.100	0.400	0.100	0.400	0.025	0.468	0.126	0.381

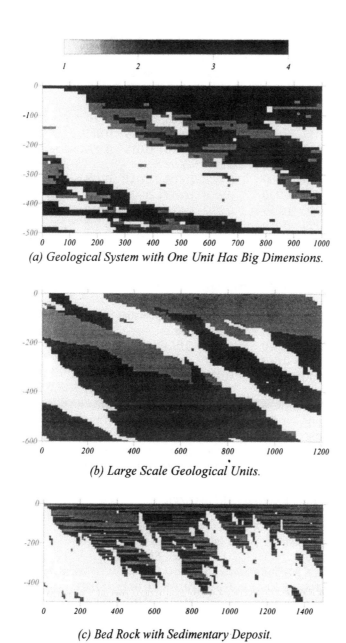

(a) Geological System with One Unit Has Big Dimensions.

(b) Large Scale Geological Units.

(c) Bed Rock with Sedimentary Deposit.

Fig.(3.6) Some Geological Patterns (Continued): (a) Four-States Formation with One Unit Possessing Relatively Large Dimensions. (b) Large Scale Geological Units. (c) Bed Rock with Sedimentary Basin.

89

Table (3.11) Statistics of Fig.(3.7a).

Sampling interval in X-axis (m)=   10.0
Sampling interval in Y-axis (m)=   10.0

Input Statistics				Calculated Statistics			
XX' - XX' Transition Probability Matrix							
0.960	0.020	0.010	0.010	0.983	0.010	0.004	0.003
0.010	0.970	0.010	0.010	0.020	0.968	0.009	0.004
0.020	0.020	0.940	0.020	0.053	0.022	0.909	0.015
0.010	0.010	0.010	0.970	0.020	0.010	0.012	0.959
YY' - YY' Transition Probability Matrix							
0.700	0.100	0.100	0.100	0.722	0.100	0.041	0.136
0.250	0.250	0.250	0.250	0.257	0.348	0.122	0.273
0.250	0.250	0.250	0.250	0.235	0.285	0.166	0.313
0.250	0.250	0.250	0.250	0.132	0.196	0.077	0.596

Table (3.12) Statistics of Fig.(3.7b).

Sampling interval in X-axis (m)=   20.0
Sampling interval in Y-axis (m)=   2.5

Input Statistics				Calculated Statistics			
XX' - XX' Transition Probability Matrix							
0.960	0.020	0.010	0.010	0.968	0.013	0.010	0.010
0.010	0.970	0.010	0.010	0.012	0.973	0.009	0.006
0.020	0.020	0.940	0.020	0.026	0.024	0.933	0.017
0.010	0.010	0.010	0.970	0.018	0.010	0.011	0.961
YY' - YY' Transition Probability Matrix							
0.500	0.200	0.200	0.100	0.422	0.321	0.120	0.138
0.250	0.250	0.250	0.250	0.268	0.312	0.164	0.255
0.250	0.250	0.250	0.250	0.206	0.234	0.209	0.351
0.250	0.250	0.250	0.250	0.206	0.299	0.184	0.311

Table (3.13) Statistics of Fig.(3.7c).

Sampling interval in X-axis (m)=   20.0
Sampling interval in Y-axis (m)=   10.0

Input Statistics				Calculated Statistics			
XX' - XX' Transition Probability Matrix							
0.700	0.100	0.100	0.100	0.650	0.098	0.072	0.180
0.100	0.600	0.200	0.100	0.126	0.673	0.093	0.109
0.200	0.200	0.500	0.100	0.181	0.288	0.323	0.209
0.100	0.200	0.200	0.500	0.120	0.223	0.059	0.598
YY' - YY' Transition Probability Matrix							
0.980	0.005	0.005	0.010	0.969	0.007	0.012	0.011
0.010	0.970	0.010	0.010	0.003	0.981	0.011	0.005
0.020	0.020	0.940	0.020	0.037	0.027	0.911	0.025
0.010	0.010	0.010	0.970	0.013	0.018	0.007	0.962

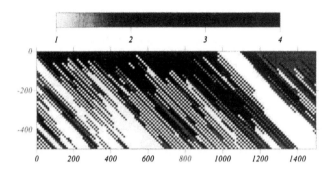

(a) Inclined Bedding with -45 degrees.

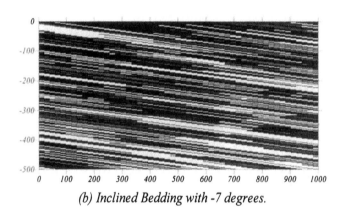

(b) Inclined Bedding with -7 degrees.

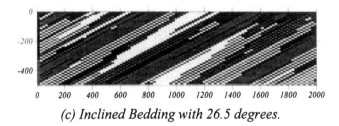

(c) Inclined Bedding with 26.5 degrees.

Fig.(3.7) Some Geological Patterns (Continued): (a) Inclined Bedding with -45°. (b) Inclined Bedding with -7°. (c) Inclined Bedding with 26.6°.

   A focus on generation of the geological structure rather than parametric variability is considered in the previous sections. An illustration of parametric variability is displayed in Fig.(3.8). Fig.(3.8a) shows a four states pattern generated according to the transition probabilities given in Table (3.14). Randomly generated hydraulic conductivity fields are introduced to the displayed pattern. In Fig.(3.8b) to each state in the system a conductivity value is assigned from a log-normal distribution with parameters given in Table (3.15). However, in Fig.(3.8c) to each state in the system a conductivity value is assigned from a uniform distribution according to data in Table (3.15).

Table(3.14) Statistics of Fig.(3.8a).

Sampling interval in X-axis (m)=			10.0				
Sampling interval in Y-axis (m)=			10.0				
Input Statistics				Calculated Statistics			
Horizontal Transition Probability Matrix							
0.700	0.100	0.100	0.100	0.682	0.113	0.095	0.110
0.200	0.600	0.100	0.100	0.214	0.599	0.077	0.111
0.100	0.050	0.800	0.050	0.085	0.044	0.818	0.053
0.100	0.150	0.250	0.500	0.083	0.180	0.261	0.477
Vertical Transition Probability Matrix							
0.250	0.250	0.250	0.250	0.263	0.192	0.406	0.139
0.250	0.250	0.250	0.250	0.264	0.196	0.400	0.140
0.250	0.250	0.250	0.250	0.268	0.180	0.413	0.139
0.250	0.250	0.250	0.250	0.292	0.181	0.382	0.145

Table (3.15) Parameters of Uniform and Log-normal Distributions.

Parameter	Uniform Distribution		Log-normal Distribution	
State	K_{min} K_{max} m/day		$<K>$ m/day	σ_K m/day
State 1	1.	5.	3.	2
State 2	5.	10.	7.5	2.5
State 3	10.	50.	30.	20.
State 4	50.	100.	75.	25.

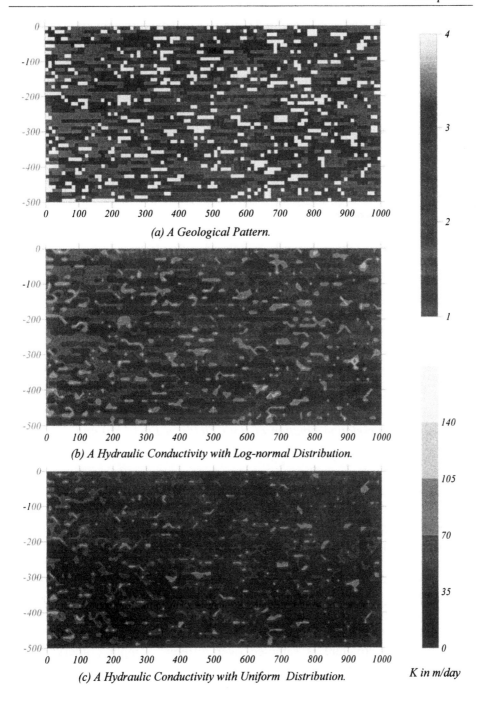

(a) A Geological Pattern.

(b) A Hydraulic Conductivity with Log-normal Distribution.

(c) A Hydraulic Conductivity with Uniform Distribution.

K in m/day

Fig.(3.8) Parametric Variability Fields.

Both figures show the same structural pattern as in the original image. The only difference is in the magnitude of each grid point. For uniform distribution the maximum value generated is 100 m/day and minimum value is 1 m/day, whereas, for lognormal distribution the maximum value generated in the field is 435 m/day and the minimum value is 0.4 m/day. The experimental autocorrelation for both fields has been calculated and displayed in Fig.(3.9). There is no significant difference between the two fields because both fields have the same structural pattern and no matter if the magnitudes are different.

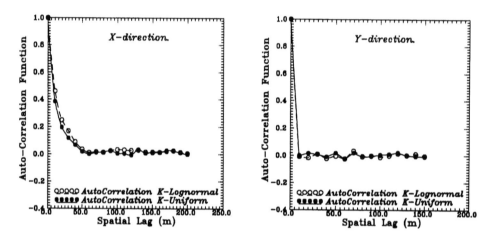

Fig.(3.9) Autocorrelation function of Both Fields in Fig.(3.8b) and Fig.(3.8c).

3.13 Hypothetical Case Study

The following steps illustrate the application of the proposed geological characterization method for field application. The procedure is presented for a hypothetical case study for demonstration purposes. Steps are illustrated below,

(*i*) A geological structure has been generated which is assumed to be the '*Real Formation*'. Fig.(3.10a) shows the '*real formation*' with dimensions 300×100 m. Four different geological materials are assumed to be present in the system which are indicated with different colours and numbered 1,2,3 and 4.

(*ii*) The given structure is sampled over different sampling intervals by superimposing a grid with a proper sampling interval. Three different sampling intervals are used. The first case $\delta x = 8$ m and $\delta y = 4$ m. The second case $\delta x = 4$ m and $\delta y = 2$ m. The third case $\delta x = 2$ m and $\delta y = 1$ m. The corresponding transitional probabilities for each case are displayed in Table (3.16), (3.17), (3.18) respectively.

Table (3.16) Statistics of Fig.(3.10b).

Sampling interval in X-axis (m)= 8.0
Sampling interval in Y-axis (m)= 4.0

Input Statistics   Calculated Statistics

Horizontal Transition Probability Matrix

0.987	0.003	0.008	0.003	0.998	0.000	0.002	0.000
0.022	0.871	0.065	0.043	0.013	0.893	0.093	0.000
0.042	0.012	0.931	0.015	0.037	0.002	0.956	0.005
0.044	0.022	0.022	0.911	0.115	0.038	0.077	0.769

Vertical Transition Probability Matrix

0.739	0.019	0.168	0.074	0.817	0.005	0.178	0.000
0.148	0.318	0.148	0.386	0.333	0.533	0.107	0.027
0.206	0.078	0.676	0.040	0.176	0.032	0.783	0.009
0.177	0.274	0.226	0.323	0.043	0.826	0.087	0.043

Table (3.17) Statistics of Fig.(3.10c).

Sampling interval in X-axis (m)= 4.0
Sampling interval in Y-axis (m)= 2.0

Input Statistics   Calculated Statistics

Horizontal Transition Probability Matrix

0.990	0.003	0.004	0.003	0.993	0.000	0.005	0.002
0.011	0.958	0.019	0.011	0.031	0.870	0.084	0.015
0.016	0.005	0.969	0.009	0.011	0.000	0.988	0.002
0.036	0.008	0.015	0.941	0.048	0.000	0.083	0.869

Vertical Transition Probability Matrix

0.813	0.011	0.090	0.086	0.852	0.002	0.146	0.001
0.119	0.689	0.042	0.150	0.244	0.565	0.092	0.099
0.095	0.043	0.840	0.023	0.046	0.011	0.940	0.003
0.081	0.347	0.257	0.314	0.037	0.333	0.284	0.346

Table (3.18) Statistics of Fig.(3.10d).

Sampling interval in X-axis (m)= 2.0
Sampling interval in Y-axis (m)= 1.0

Input Statistics   Calculated Statistics

Horizontal Transition Probability Matrix

0.995	0.001	0.002	0.001	0.997	0.000	0.002	0.000
0.006	0.979	0.009	0.006	0.072	0.893	0.023	0.012
0.008	0.003	0.985	0.004	0.009	0.000	0.990	0.001
0.019	0.003	0.008	0.970	0.025	0.004	0.032	0.939

Vertical Transition Probability Matrix

0.906	0.005	0.045	0.043	0.945	0.002	0.051	0.001
0.059	0.845	0.021	0.075	0.107	0.768	0.052	0.072
0.047	0.021	0.920	0.011	0.060	0.003	0.935	0.002
0.040	0.174	0.129	0.657	0.135	0.161	0.109	0.596

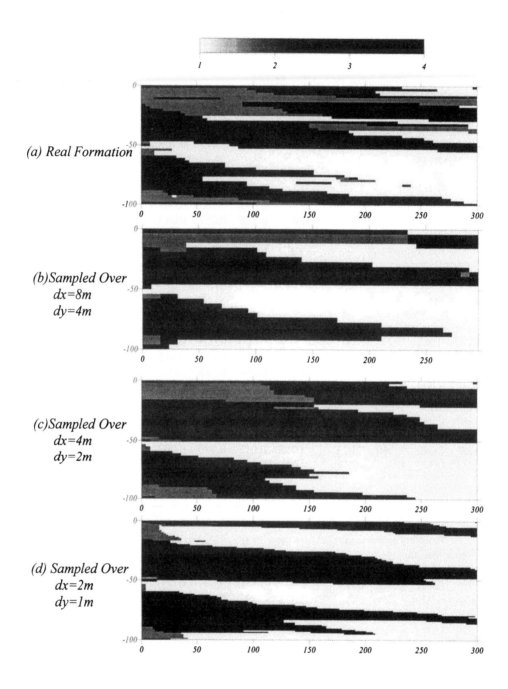

Fig.(3.10) 'Real Formation' Compared with Some Synthetic Realizations Sampled over Different Sampling Intervals.

(*iii*) During the process of estimating transition probabilities some kind of averaging or upscaling should be done since in large sampling intervals one may have different geological materials in one cell.

(*iv*) After estimating these transition probabilities with the chosen intervals, simulation step is performed using the developed computer programme GEOSIM1.

(*v*) Fig.(3.10 b,c &d) shows three single realizations generated with the estimated transition probabilities using the algorithm presented in section 3.7.

In this case study the realizations are conditioned on both the top and left boundary to minimize uncertainty in the generated images and to show how conditioning can be implemented. It is clear from this case that the three realizations look quite close to the '*real formation*', but realization *c* seems the best.

3.14 Summary and Conclusions

A stochastic methodology for characterizing geological formations using soft information has been presented. The heterogeneity and anisotropy, which would represent the stratified character observed in sedimentary rocks, have been modelled with two transition probability matrices. A horizontal transition probability matrix describes the variation in the geological materials in horizontal direction, and a vertical transition probability matrix describes the variations in the vertical direction. Transition probabilities are used to calculate the conditional probabilities in the decomposition formula Eq.(3.12). Two computer models GEOSIM1, GEOSIM2 have been developed to implement the proposed characterization method. The required input for the programs includes the dimensions of the geological section (length and depth), the number of the geological units present in the system, transition probabilities and sampling intervals over which these transitions are estimated.

The following conclusions and remarks can be made:

(1) The present work demonstrates that the current stochastic methodology is capable of characterizing different patterns of natural geological formations, stratifications, layering systems or perfectly and imperfectly stratified systems, inclusions, channelling, horizontal and inclined bedding. The results are comparable to what is expected to be found in natural formations based on geological experience and common sense.

(2) The present technique is more general than the one used by Krumbein [1967] which synthesizes a geological stratigraphic sequence using a single transition probability matrix.

(3) The technique calls for only a few simulation parameters (transition probabilities) which can be estimated easily.

(4) The method reproduces pattern statistics. The given and simulated statistics show acceptable agreements with only minor fluctuations in some cases. These are due to the use of a large number of states in a small bounded domain. The discrepancies are attributed to the small number of transitions of one or more states existing in the geological system.

(5) Natural extension of this algorithm to three dimensions is quite simple, just one extra transition probability matrix will be used to account for the transitions in the third direction.

(6) This methodology can be linked with one of the available techniques in the literature such as the turning bands method to produce a more detailed characterization. This part has been carried out in Chapter 5.

(7) The methodology has been tested on an artificially generated geological structure (hypothetical case) and proven fairly successful.

(8) It has been shown that the methodology is not only capable of generating geological structures but also parametric variability with different parametric distributions. In this case the spatial structure of the parameters is characterized by conditional probabilities and not by auto-covariances.

(9) The limitations of the present methodology consists in three points. The first is the distinction of the different lithologies used in establishing the transition probability matrices. The second is that the number of transitions measured should be sufficiently large to obtain statistically reliable estimates of the transition probabilities. The third is that the method relies on first-order and stationary transitions. However, an extension can be made to higher-order and nonstationary transtions but more data will be needed.

(10) Another limitation is the fact that the historical sequence of geological deposits is not necessarily Markovian. One can improve the result by omitting unrealistic realizations. This is another way of incorporating '*soft data*', i.e. geological knowledge. This aspect is further elaborated in Chapter 7.

CHAPTER 4

Flow in Heterogeneous Formations Characterized by a Coupled Markov Chain Model

4.1 Introduction

Significant progress of groundwater flow and physico-chemical transport in stationary Gaussian random fields has been achieved. Many researchers [e.g. Smith et al., 1979b and Ababou et al., 1989] show still vivid interest in methods suitable to describe the hydrodynamics of flow in heterogeneous field. Only a limited number of these studies is devoted to the use of geologically-based stochastic characterization methods. Just recently, in the hydrogeological community, a considerable attention on evaluating the effects of deviations from the assumption of Gaussian fields is arising. Therefore, in this chapter an attempt is made to study Markovian fields.

The goal of this chapter is to investigate the hydrodynamics of flow in approximate geological structures characterized by a coupled Markov chain model presented in the previous chapter. A finite difference numerical model of flow in terms of potentials and stream functions has been developed. The effects of geological variability on the hydraulic head and specific discharge are demonstrated. Furthermore, the resulting variability in the flow velocities which is the cause of spreading and dilution of the solute plumes is explained later in Chapter 6. A steady saturated confined groundwater flow problem is investigated. Hydraulic anisotropy of the formation structure is shown with the present model. The probability density function of the effective hydraulic conductivity is inferred by performing some numerical experiments.

4.2 Procedure for Flow Simulation by The Coupled Markov Chain Model

The simulation procedure involves the following steps:

Step 1. Data Collection: Well logs, bore holes, geological maps, all existing information on the geological structure of the site is collected from field survey.

Step 2. Statistical Inferences: At this stage, identification of the categories of geological materials (e.g., sand, clay,...etc.) present in the aquifer system is done. These are referred to in the following as states. Well logs and geological maps are discretized with proper sampling intervals, the transition probability matrices are determined with respect to the chosen sampling intervals.

Step 3. Geological Simulation: The number of states in the geological system, transition probabilities, domain dimensions, and sampling intervals are used to generate several realizations of the geological structure.

Step 4. Geological Judgement: In case of single realization approach, geologists experience, technical background of practitioners, intuition and common sense should be involved to decide for the most probable image of the geological pattern from the several generated realizations. However, in case of multi-realizations approach, Monte-Carlo technique is applied to estimate the degree of uncertainty of output variables given the uncertainty in the input parameters. This approach is often used for risk assessment which is considered in more detail in Chapter 7.

Step 5. Parameter Assignation: At this level, the parameter in each state (geological material) is assigned a value. This parameter may refer to hydraulic conductivity, porosity, dispersivity, etc. These values can be obtained either from the literature or empirically from the grain size distribution of the samples or any source of soft information.

Step 6. Flow Simulator: The maps of the hydrogeological parameters calculated by step 5 are used as input for the flow simulation together with initial and boundary conditions. Step 1, 2, 3 and 4 are discussed in details in Chapter 3 [see also Elfeki, 1994], a brief review is given in section 4.3. The remaining sections discuss step 6.

4.3 Modelling The Heterogeneous Structure

The two-dimensional Markov model developed in Chapter 3 is used to generate several realizations of the geological structures. These realizations are characterized by soft information (i.e. in terms of transition probabilities) rather than hard data. In this study each geological material will be idealized as one unit having homogeneous properties.

4.4 Governing Equations of Flow Problem

4.4.1 Continuous Form of Flow in Terms of Hydraulic Head

For steady state two-dimensional (in vertical plane or in cross-section) saturated incompressible fluid flow in an anisotropic heterogeneous confined aquifer, the governing equation in the absence of source and sink terms is:

$$\frac{\partial}{\partial x}\left(K_{xx}\frac{\partial \Phi}{\partial x} \right) + \frac{\partial}{\partial y}\left(K_{yy}\frac{\partial \Phi}{\partial y} \right) = 0 \qquad \in \Omega \qquad (4.1)$$

where, K_{xx} is hydraulic conductivity in x-direction, K_{yy} is hydraulic conductivity in y-direction, Φ is hydraulic or piezometric head, and Ω is domain of interest.

No-flow (Neumann condition) or constant head (Dirichlet condition) are specified on the boundaries of the flow domain, that is,

$$\frac{\partial}{\partial n}[\Phi(x,y)] = 0 \qquad on \quad x,y \in \Gamma_1 \qquad (4.2)$$

$$\Phi(x,y) = \Phi_o \qquad on \quad x,y \in \Gamma_2 \qquad (4.3)$$

where, Γ is boundary of the domain, $\Gamma = \Gamma_1 + \Gamma_2$, n is the unit vector normal to the boundary pointing outward, and Φ_o is the prescribed head.

4.4.2 Continuous Form of Flow in Terms of Stream Function

It is sometimes desirable to present the pattern of streamlines. There are two methods to obtain the streamlines. In the first method the velocity vectors are determined from the potential solution by determination of the local potential gradients and the local hydraulic conductivity. When the velocity vectors are obtained at a number of locations, the streamlines can be plotted graphically by drawing tangents to these vectors in the flow directions or numerically be particle tracking algorithm. A more accurate method, which requires additional computational effort is solving the equation of the stream function. The differential equation for the stream function Ψ can be written [for more details and derivations see Akker, 1982] as,

$$\frac{\partial}{\partial x}\left(\frac{1}{K_{xx}}\frac{\partial \Psi}{\partial x} \right) + \frac{\partial}{\partial y}\left(\frac{1}{K_{yy}}\frac{\partial \Psi}{\partial y} \right) = 0 \qquad \in \Omega \qquad (4.4)$$

with boundary conditions defined as,

$$\frac{\partial}{\partial n}[\Psi(x,y)] = 0 \qquad on \quad x,y \in \Gamma_1 \qquad (4.5)$$

$$\Psi(x,y) = \Psi_o \qquad on \quad x,y \in \Gamma_2 \qquad (4.6)$$

where, Ψ_o is the prescribed stream function.

4.5 Spatial Discrtization and Finite Difference Formulation

4.5.1 Spatial Discretization of Flow Field

Solving the governing differential equations over the domain of interest Ω requires a spatial discretization of the flow field. It is necessary to discretize the domain Ω into a number of cells where the variable of interest is defined in its centroid. These centroids are known as grid points or nodes. Fig.(4.1) illustrates a domain discretization. For accurate flow simulation the following resolutionrequirements must be fulfilled,

$$\Delta x \leq \delta x$$
$$\Delta y \leq \delta y \qquad (4.7)$$

where, Δx and Δy are domain discretization in x- and y-direction respectively, and δx and δy are sampling intervals of the geological model in x- and y-direction respectively.

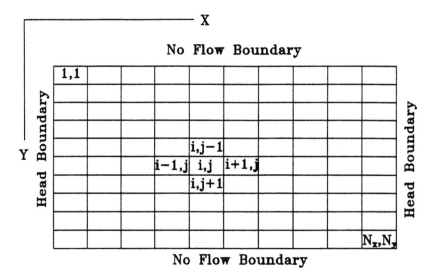

Fig.(4.1) Domain Discretization, Numerical Operator, and Boundary Conditions.

The flow domain is discretized into a number of nodes that is equal to the number of cells used in the geological model. That means Eq.(4.7) transforms into an equality.

4.5.2 Discrete Form of The Hydraulic Head Field

A finite difference model has been developed for discretization of Eq.(4.1). A numerical scheme with a five-points operator shown in Fig.(4.1) is used. The finite difference analog for the derivatives are given in the following expressions,

$$K_{xx}\left(\frac{\partial \Phi}{\partial x}\right) \approx K_{xx_{i+1/2,j}}\left[\frac{\Phi_{i+1,j} - \Phi_{i,j}}{\Delta x}\right] \qquad (4.8)$$

where, $K_{xxi+1/2,j}$ is the interface hydraulic conductivity between node $(i+1,j)$ and node (i,j). This hydraulic conductivity could be estimated with arithmetic mean, harmonic mean or geometric mean of the surrounding nodes in x-direction. The arithmetic mean is used [Kinzelbach, 1986; Bear and Verruijt, 1987],

$$K_{xx_{i+1/2,j}} = \frac{1}{2}\left[K_{xx_{i+1,j}} + K_{xx_{i,j}}\right] \qquad (4.9)$$

Similarly,

$$K_{yy}\left(\frac{\partial \Phi}{\partial y}\right) \approx K_{yy_{i,j+1/2}}\left[\frac{\Phi_{i,j+1} - \Phi_{i,j}}{\Delta y}\right] \qquad (4.10)$$

with $K_{yyi,j+1/2}$ is given by

$$K_{yy_{i,j+1/2}} = \frac{1}{2}\left[K_{yy_{i,j+1}} + K_{yy_{i,j}}\right] \qquad (4.11)$$

Further evaluation leads to

$$\frac{\partial}{\partial x}\left(K_{xx}\frac{\partial \Phi}{\partial x} \right) \approx \frac{K_{xx_{i+1/2,j}}\left[\dfrac{\Phi_{i+1,j} - \Phi_{i,j}}{\Delta x}\right] - K_{xx_{i-1/2,j}}\left[\dfrac{\Phi_{i,j} - \Phi_{i-1,j}}{\Delta x}\right]}{\Delta x} \quad (4.12)$$

$$\frac{\partial}{\partial y}\left(K_{yy}\frac{\partial \Phi}{\partial y} \right) \approx \frac{K_{yy_{i,j+1/2}}\left[\dfrac{\Phi_{i,j+1} - \Phi_{i,j}}{\Delta y}\right] - K_{yy_{i,j-1/2}}\left[\dfrac{\Phi_{i,j} - \Phi_{i,j-1}}{\Delta y}\right]}{\Delta y} \quad (4.13)$$

substitution of Eq.(4.12) and Eq.(4.13) into Eq.(4.1) leads to the finite difference analog for the partial differential equation Eq.(4.1) as,

$$A_{i,j}\Phi_{i+1,j} + B_{i,j}\Phi_{i,j-1} + C_{i,j}\Phi_{i-1,j} + D_{i,j}\Phi_{i,j+1} - E_{i,j}\Phi_{i,j} = 0$$

where,

$$A_{i,j} = K_{xx_{i+1/2,j}} / \Delta x^2$$

$$B_{i,j} = K_{yy_{i,j+1/2}} / \Delta y^2$$

$$C_{i,j} = K_{yy_{i,j-1/2}} / \Delta x^2 \quad (4.14)$$

$$D_{i,j} = K_{yy_{i,j-1/2}} / \Delta y^2$$

$$E_{i,j} = A_{i,j} + B_{i,j} + C_{i,j} + D_{i,j}$$

4.5.3 Discrete Form of Velocity Field

After the solution of the flow equation to calculate the potential head distribution it is straightforward to estimate the gradient field and subsequently the Darcy's velocity field on the grid. This is can be done by differentiation as in the following formulas given by,

$$q_{x_{i+1/2,j}} = - K_{xx}\left(\frac{\partial \Phi}{\partial x}\right) \approx - K_{xx_{i+1/2,j}}\left[\frac{\Phi_{i+1,j} - \Phi_{i,j}}{\Delta x}\right] \quad (4.15)$$

$$q_{y_{i,j+1/2}} = -K_{yy}\left(\frac{\partial \Phi}{\partial y}\right) \approx -K_{yy_{i,j+1/2}}\left[\frac{\Phi_{i,j+1} - \Phi_{i,j}}{\Delta y}\right] \qquad (4.16)$$

where, $q_{xi+1/2,j}$, $q_{yi,j+1}$ are the internodal Darcy's velocity components between nodes (i,j) and $(i+1,j)$, and between nodes (i,j) and $(i,j+1)$.

From the Darcy's velocities the pore-velocities are calculated by dividing Eqs.(4.15) and (4.16) by the effective porosity of the medium. This is essential to the transport models that will be presented in Chapter 6. The accuracy of using differentials to obtain the velocity field introduces relatively large numerical errors, because of the low degree of interpolation functions used (see Eq. 4.12 and 4.13). Therefore, the stream function is considered.

4.5.4 Discrete Form of Stream Function Field

A discrete form of stream function can be derived in a similar way to what has been done to the hydraulic head. The finite difference analog for the partial differential equation Eq.(4.4) is,

$$\acute{A}_{i,j}\Psi_{i+1,j} + \acute{B}_{i,j}\Psi_{i,j-1} + \acute{C}_{i,j}\Psi_{i-1,j} + \acute{D}_{i,j}\Psi_{i,j+1} - \acute{E}_{i,j}\Psi_{i,j} = 0$$

where,

$$\acute{A}_{i,j} = \left(1 \,/\, K_{xx_{i+1/2,j}}\right) /\Delta x^2$$

$$\acute{B}_{i,j} = \left(1 \,/\, K_{yy_{i,j+1/2}}\right) /\Delta y^2$$

$$\acute{C}_{i,j} = \left(1 \,/\, K_{yy_{i,j-1/2}}\right) /\Delta x^2 \qquad (4.17)$$

$$\acute{D}_{i,j} = \left(1 \,/\, K_{yy_{i,j-1/2}}\right) /\Delta y^2$$

$$\acute{E}_{i,j} = \acute{A}_{i,j} + \acute{B}_{i,j} + \acute{C}_{i,j} + \acute{D}_{i,j}$$

4.6 Solution of Field Equations

A large number of solvers are available for systems of linear equations and some of the efficient solvers, in case of large number of nodes, are the iterative ones. All the iterative solvers start with an initial guess of the field variable and in each iteration a new and better approximation is computed. It has been proven that the method of conjugate gradient is powerful in addressing highly heterogeneous medium. This method is used in the current study. The formulas and the algorithm for implementation are presented. The algorithm used here is the one by [Strikwerda, 1989]. Some modifications are adopted to handle the heterogeneity of the medium reflected in the coefficients *A*, *B*, *C*, *D* and *E*. The equations to solve are in the form of Eq.(4.14) or Eq.(4.17) which form the linear system $ax = b$ where, a is positive definite matrix and the vector b contains both zeros and the values of the solution on the boundary. The procedure involves the following steps. The notation used in this procedure are ^Xxe hydraulic head notations and it can be easily changed for the stream function.

First step: an initial iterate $\Phi^o_{i,j}$ is given and then the residual $r^o_{i,j}$ is computed as,

$$r^o_{i,j} = A_{i,j}\Phi^o_{i+1,j} + B_{i,j}\Phi^o_{i,j-1} + C_{i,j}\Phi^o_{i-1,j} + D_{i,j}\Phi^o_{i,j+1} - E_{i,j}\Phi^o_{i,j} \quad (4.18)$$

where a matrix $P^o_{i,j}$ is introduced as

$$P^o_{i,j} = r^o_{i,j} \quad (4.19)$$

with $|r^o|^2$ also being computed by accumulating the products $r^o_{i,j}\, r^o_{i,j}$. In mathematical form,

$$| r^o |^2 = \sum \left[r^o_{i,j}\right]^2 \qquad for\ all\ i,j \quad (4.20)$$

Another matrix $q_{i,j}$ is introduced and computed as

$$q^o_{i,j} = - A_{i,j}r^o_{i+1,j} - B_{i,j}r^o_{i,j-1} - C_{i,j}r^o_{i-1,j} - D_{i,j}r^o_{i,j+1} + E_{i,j}r^o_{i,j} \quad (4.21)$$

and the inner product (P^o, q^o), is computed by accumulating the products $P^o_{i,j}\, q^o_{i,j}$ to evaluate the parameter α_o as,

$$\alpha_o = \frac{| r^o |^2}{(P^o, q^o)} \quad (4.22)$$

Note that for Dirichlet boundary condition (prescribed head boundary) r^k, P^k, and q^k where k denotes the iteration number, should be zero on the boundary.

Second step: begin the main computation loop. $\Phi_{i,j}$ and $r_{i,j}$ are updated by

$$\Phi_{i,j}^{k+1} = \Phi_{i,j}^{k} + \alpha_k P_{i,j}^{k}$$

$$r_{i,j}^{k+1} = r_{i,j}^{k} - \alpha_k q_{i,j}^{k} \tag{4.23}$$

with $| r^{k+1} |^2$ also computed. Another parameter β_k is computed by the formula,

$$\beta_k = \frac{| r^{k+1} |^2}{| r^k |^2} \tag{4.24}$$

then P and q are updated by

$$P_{i,j}^{k+1} = r_{i,j}^{k+1} + \beta_k P_{i,j}^{k} \tag{4.25}$$

$$q_{i,j}^{k+1} = \left[E_{i,j} r_{i,j}^{k+1} - A_{i,j} r_{i+1,j}^{k+1} - B_{i,j} r_{i,j-1}^{k+1} - C_{i,j} r_{i-1,j}^{k+1} - D_{i,j} r_{i,j+1}^{k+1} \right] + \beta_k q_{i,j}^{k}$$

and the inner product (P^{k+1}, q^{k+1}) is computed.

Third step: α_{k+1} is computed as the ratio,

$$\alpha_{k+1} = \frac{| r^{k+1} |^2}{(P^{k+1}, q^{k+1})} \tag{4.26}$$

and k is incremented.

The conjugate gradient method is terminated when $| r^k |$ is sufficiently small. As with the general iterative methods, the method should be continued until the error in the iteration is comparable to the truncation error in the numerical scheme.

4.7 Computational Algorithm for Flow Simulation

A single realization approach is followed to simulate the flow behaviour in two dimensional heterogeneous geological structures. The computations start as follows. A grid is superimposed over the geological image. Each nodal point is assigned a hydraulic conductivity value according to the corresponding geological material at that node. Table (4.1) shows the corresponding hydraulic conductivity values of each geological unit. The boundary conditions are applied to the domain. Two types of boundary conditions are used, prescribed head and no flow boundaries. Horizontal flow from left to right has been simulated which is described by prescribed head boundary on the left and right side, and no flow boundary at top and bottom of the flow domain as shown in Fig.(4.1). For head calculations the left boundary is given a value of one ($\Phi_o = 1$) and the right boundary is zero ($\Phi_{Lx} = 0$). On the other hand, the solution for the stream function is done by inverting the hydraulic conductivity field and the boundary conditions i.e. the prescribed constant head boundary is considered as a no-flow boundary (the variation of this stream function across the boundary is zero) on the left and right boundaries of the flow domain. No-flow boundary is inverted to a prescribed constant stream function boundary. The value of the stream function is zero on the top boundary ($\Psi_o = 0$), and one on the bottom boundary ($\Psi_{Ly} = 1$). The procedure for the solution starts by initializing the response variable at each node on the grid to zero. The solution procedure stops after a reasonable number of iterations when the solution converges to the true solution.

Table(4.1) Hydraulic Conductivities of the States used in Simulation Program.

State	Colour on Map	Isotropic Hydraulic Conductivity
1	White	100. m/day
2	Light Grey	10. m/day
3	Dark Grey	1. m/day
4	Black	0.1 m/day

4.8 Verifications and Some Applications of The Simulation Model

Fig.(4.2) shows the flow problem in terms of potentials and stream functions. Some geological structures with realistic characteristics are generated. The statistical characterization of these structures are displayed in Tables (4.2), (4.3) and (4.4). The corresponding flow problems are solved. Fig.(4.3) displays flow pattern in a large scale geological formation. Fig.(4.4) displays flow pattern in a thin layered formation with inclination of -7° to the horizontal. Fig.(4.5) displays a flow pattern in a horizontal stratified aquifer.

The results of this model can be verified by checking three conditions. First, the orthogonality of the flowlines and the equipotential lines in case of isotropic cell hydraulic conductivity must be achieved. Second, the water particles must use the shortest path-lowest resistance i.e. the flowlines should follow the regions of high conductivity values and take the shortest path in low conductivity regions. Third, the mass conservation principle must apply, which means that the quantity of seepage calculated at any vertical section must be the same. In the present model the quantity of seepage is calculated at two sections on the middle of the formation (i.e. $L_x/2$) and at a section on the first third (i.e. $L_x/3$). The quantity of seepage for the three applications are calculated and tabulated in Table (4.5). The results of the present model are fulfilling the three conditions fairly good.

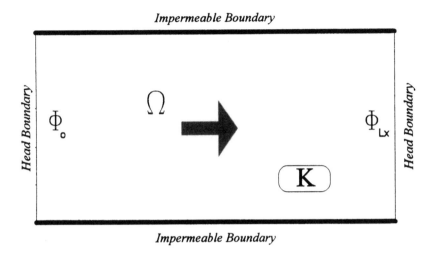

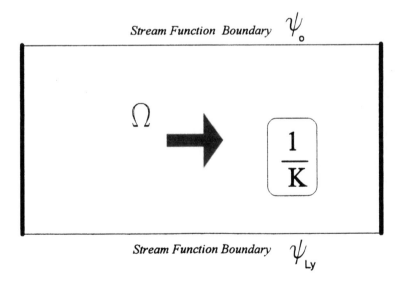

Fig.(4.2) Flow Problem in Terms of Potentials and Stream Functions.

Table (4.2) Statistics of Fig.(4.3).

Sampling interval in X-axis (m)=   1.0
Sampling interval in Y-axis (m)=   0.5

Input Statistics   Calculated Statistics

Horizontal Transition Probability Matrix

State	1	2	3	4	State	1	2	3	4
1	0.900	0.030	0.030	0.040	1	0.924	0.017	0.047	0.013
2	0.010	0.970	0.010	0.010	2	0.022	0.960	0.016	0.002
3	0.010	0.020	0.960	0.010	3	0.024	0.010	0.963	0.003
4	0.040	0.040	0.010	0.910	4	0.093	0.053	0.099	0.755

Vertical Transition Probability Matrix

State	1	2	3	4	State	1	2	3	4
1	0.970	0.010	0.010	0.010	1	0.939	0.020	0.038	0.004
2	0.040	0.900	0.030	0.030	2	0.022	0.940	0.033	0.006
3	0.020	0.010	0.960	0.010	3	0.017	0.007	0.972	0.005
4	0.040	0.040	0.010	0.910	4	0.134	0.043	0.062	0.761

Table (4.3) Statistics of Fig.(4.4).

Sampling interval in X-axis (m)=   2.0
Sampling interval in Y-axis (m) =   0.25

Input Statistics   Calculated Statistics

XX' - XX' Transition Probability Matrix

State	1	2	3	4	State	1	2	3	4
1	0.960	0.020	0.010	0.010	1	0.958	0.022	0.011	0.009
2	0.010	0.970	0.010	0.010	2	0.009	0.973	0.010	0.008
3	0.020	0.020	0.940	0.020	3	0.024	0.015	0.939	0.023
4	0.010	0.010	0.010	0.970	4	0.011	0.008	0.012	0.969

YY' - YY' Transition Probability Matrix

State	1	2	3	4	State	1	2	3	4
1	0.250	0.250	0.250	0.250	1	0.252	0.347	0.151	0.250
2	0.250	0.250	0.250	0.250	2	0.186	0.310	0.188	0.316
3	0.250	0.250	0.250	0.250	3	0.194	0.302	0.187	0.317
4	0.250	0.250	0.250	0.250	4	0.206	0.258	0.173	0.362

Table (4.4) Statistics of Fig.(4.5).

Sampling interval in X-axis (m)=   1.0
Sampling interval in Y-axis (m)=   1.0

Input Statistics   Calculated Statistics

Horizontal Transition Probability Matrix

State	1	2	3	4	State	1	2	3	4
1	0.960	0.020	0.010	0.010	1	0.961	0.029	0.006	0.005
2	0.010	0.970	0.010	0.010	2	0.009	0.977	0.009	0.005
3	0.020	0.020	0.940	0.020	3	0.032	0.031	0.928	0.009
4	0.010	0.010	0.010	0.970	4	0.020	0.011	0.011	0.957

Vertical Transition Probability Matrix

State	1	2	3	4	State	1	2	3	4
1	0.500	0.250	0.150	0.100	1	0.331	0.433	0.120	0.115
2	0.250	0.500	0.150	0.100	2	0.202	0.626	0.084	0.089
3	0.250	0.150	0.500	0.100	3	0.248	0.254	0.382	0.116
4	0.250	0.150	0.100	0.500	4	0.342	0.185	0.025	0.447

Table (4.5) Quantity of Seepage Calculated from The Model.

Application No.	Seepage at $L_x/2$	Seepage at $L_x/3$	Abs.Error (m^2/day)
1	4.524 m^2/day	4.552 m^2/day	0.028
2	10.537 m^2/day	10.509 m^2/day	0.028
3	12.667 m^2/day	12.613 m^2/day	0.054

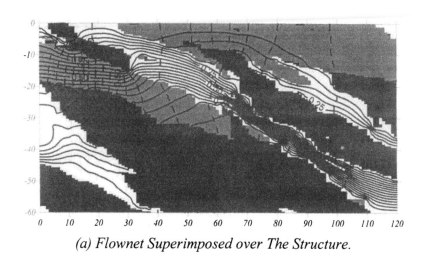

(a) Flownet Superimposed over The Structure.

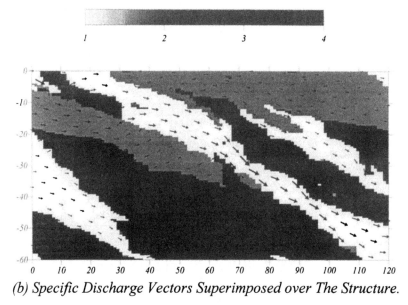

(b) Specific Discharge Vectors Superimposed over The Structure.

Fig.(4.3) Flow Simulation in Large Scale Geological Units.

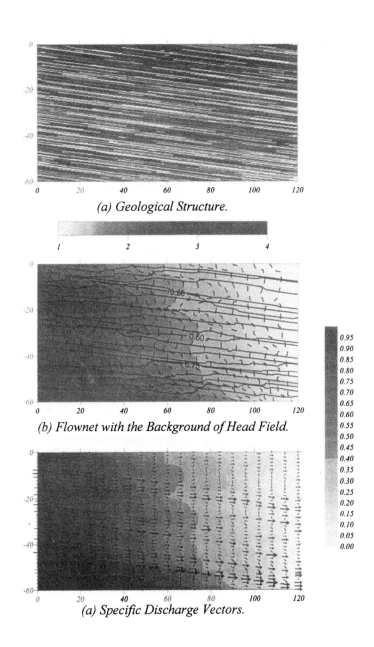

(a) Geological Structure.

(b) Flownet with the Background of Head Field.

(a) Specific Discharge Vectors.

Fig.(4.4) Flow Simulation in Inclined Bedding with 7°.

113

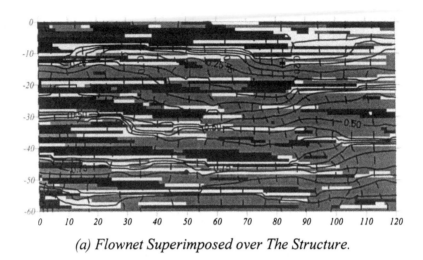

(a) Flownet Superimposed over The Structure.

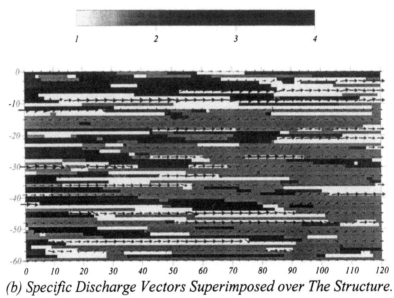

(b) Specific Discharge Vectors Superimposed over The Structure.

Fig.(4.5) Flow Simulation in Stratified Formation.

114

4.9 Numerical Experiments on Effective Hydraulic Conductivity

4.9.1 Purpose of The Experiments

Conductivity of saturated flow in heterogeneous medium may be described macroscopically through the effective conductivity, defined as, the global hydraulic conductivity of a homogeneous sample that has the same discharge rate as the heterogeneous sample. The macroscopic conductivity is often estimated in the laboratory by column experiments on undisturbed and disturbed geological samples taken from the field. In many practical situations, there is often insufficient samples collected from the field. Therefore, numerical experiments may be designed to replace laboratory experiments to estimate the effective or representative conductivity and its frequency distribution throughout the aquifer. Samples, which are taken from the same environment and having the same composition and dispositional pattern, are assumed to represent realizations of a spatial random function. Monte-Carlo approach is followed to perform these numerical experiments by generating several realizations of sample compositions and its micro-structures. Table (4.6) displays the analogy between laboratory and numerical experiments.

4.9.2 Geological Samples

Two types of structures are considered in the experimental program. Horizontal stratifications and inclined bedding are generated. The geological structure of both patterns possesses the same statistical properties in terms of transition probabilities. The probabilities used to generate the samples are displayed in Table (4.7) and Table (4.8) for horizontal stratifications and inclined bedding respectively. Single realizations of geological patterns are displayed in Fig.(4.6).

4.9.3 Design of The Experiments

Monte-Carlo analysis relies on repetitive generation of replica of hydrogeological parameters and subsequent solution of the set of deterministic flow problems. In the present case, micro-structures or laminations with specified conductivities are generated with stochastic description given in terms of transition probabilities. Two-dimensional flow experiments are carried out. The test cross-section is square in shape ($L_x = L_y$) or the aspect ratio (L_x/L_y) is 1. A uniform grid is used with 10,000 cells and with cell dimensions ($\Delta x = \Delta y = 1$m). The boundary conditions are specified to be one on the left (on the top) and zero on the right (on the bottom) in case of horizontal flow and (vertical flow) respectively. The first approximation of the head in the domain is zero everywhere for all experiments. A large number of 1600 numerical experiments has been carried out. The hydraulic conductivity fields are generated by assigning each geological material a hydraulic conductivity

Table (4.6) Analogy between Laboratory and Numerical Experiments.

Exp.Type Item of Analogy	Laboratory Experiments	Numerical Experiments
(1) Geological Samples	Realizations of geological samples are obtained from the real aquifer at different locations. A large number of samples is required to get a meaningful statistics.	Realizations of the geological samples are generated using Markov model. The statistics of the generated realizations are similar to the statistics of the real aquifer in terms of transition probabilities.
(2) Setup	An experimental setup is built to estimate the effective properties of the geological samples.	A computer model is built to estimate the effective properties of the generated realizations which have the same statistics as the real ones.
(3)Groundwater Flow	Water is allowed to flow through the samples under known gradient between the upstream and downstream sections of the samples, and the quantity of seepage is measured.	Numerical flow model is constructed to solve the governing equations to estimate the groundwater head in the flow domain, and the quantity of seepage is calculated.
(4)Performing The Experiments	The experiments are repeated several times on all the available samples to estimate mean, standard deviation and pdf of the effective conductivity.	Monte-Carlo approach is followed to estimate all the statistical properties: mean, standard deviation, and pdf of effective conductivity.

Table(4.7) Statistics of Fig.(4.6a).

Sampling interval in X-axis		=		1 *Cell size*				
Sampling interval in Y-axis		=		1 *Cell size*				

Input Statistics   Calculated Statistics

Horizontal Transition Probability Matrix

State	1	2	3	4	State	1	2	3	4
1	0.960	0.020	0.010	0.010	1	0.961	0.018	0.012	0.009
2	0.010	0.970	0.010	0.010	2	0.012	0.967	0.008	0.012
3	0.020	0.020	0.940	0.020	3	0.017	0.018	0.944	0.021
4	0.010	0.010	0.010	0.970	4	0.012	0.012	0.011	0.965

Vertical Transition Probability Matrix

State	1	2	3	4	State	1	2	3	4
1	0.250	0.250	0.250	0.250	1	0.237	0.286	0.167	0.310
2	0.250	0.250	0.250	0.250	2	0.186	0.339	0.174	0.301
3	0.250	0.250	0.250	0.250	3	0.252	0.330	0.157	0.261
4	0.250	0.250	0.250	0.250	4	0.282	0.266	0.185	0.267

Table (4.8) Statistics of Fig.(4.6b).

Sampling interval in X-axis		=		1 *Cell size*				
Sampling interval in Y-axis		=		1 *Cell size*				

Input Statistics   Calculated Statistics

XX' - XX' Transition Probability Matrix

State	1	2	3	4	State	1	2	3	4
1	0.960	0.020	0.010	0.010	1	0.964	0.020	0.009	0.007
2	0.010	0.970	0.010	0.010	2	0.010	0.970	0.009	0.012
3	0.020	0.020	0.940	0.020	3	0.020	0.021	0.937	0.022
4	0.010	0.010	0.010	0.970	4	0.009	0.009	0.012	0.969

YY' - YY' Transition Probability Matrix

State	1	2	3	4	State	1	2	3	4
1	0.250	0.250	0.250	0.250	1	0.304	0.233	0.148	0.315
2	0.250	0.250	0.250	0.250	2	0.162	0.389	0.155	0.293
3	0.250	0.250	0.250	0.250	3	0.229	0.329	0.126	0.316
4	0.250	0.250	0.250	0.250	4	0.246	0.203	0.168	0.383

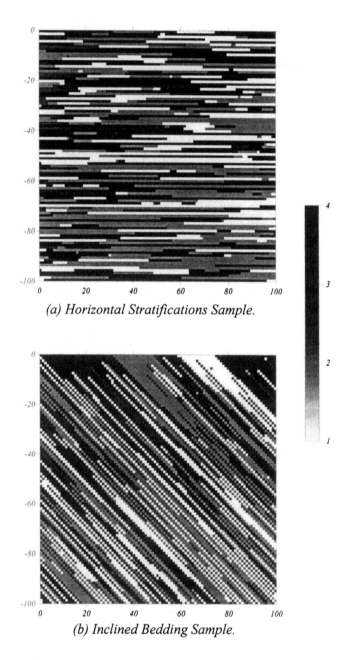

(a) Horizontal Stratifications Sample.

(b) Inclined Bedding Sample.

Fig.(4.6) Single Realization of The Geological Samples Used in The Experiments. (a) Horizontal Stratifications. (b) Inclined Bedding.

118

value. Two hypothetical case studies are investigated. The first is designed to study relatively low contrast in the conductivity. The second is designed to study a relatively high contrast. Tables (4.9) and (4.10) display the hydraulic conductivity values corresponding to each state in both case studies respectively.

From Table (4.9) and Table (4.10) one can estimate the variability of the formations in both cases. which is in the range of 20 m/day to 80 m/day in case 1 and in the range of 0.01 m/day to 100. m/day in case 2. In case 2 the system of equations are ill-conditioned which needs a reasonable amounts of computer time to get a satisfactory convergence. In all the experiments the convergence criterion is 0.001 and the maximum number of iterations used is 5000.

Table (4.9) Conductivities in Case Study No. 1 'Low Contrast'.

State	Colour on The Map	Isotropic Hydraulic Conductivity
1	White	80.0 m/day
2	Light Grey	60.0 m/day
3	Dark Grey	40.0 m/day
4	Black	20.0 m/day

Table(4.10) Conductivities in Case Study No. 2 'High Contrast'.

State	Colour on The Map	Isotropic Hydraulic Conductivity
1	White	100.0 m/day
2	Light Grey	10.00 m/day
3	Dark Grey	1.000 m/day
4	Black	0.010 m/day

4.9.4 Flow Scenarios

In performing the numerical experiments, four scenarios are invoked. These scenarios are displayed in Table (4.11). The head distributions are presented in Fig.(4.7) for horizontal stratifications and Fig.(4.8) for inclined bedding formation under both ˙ ₎w and high contrasts.

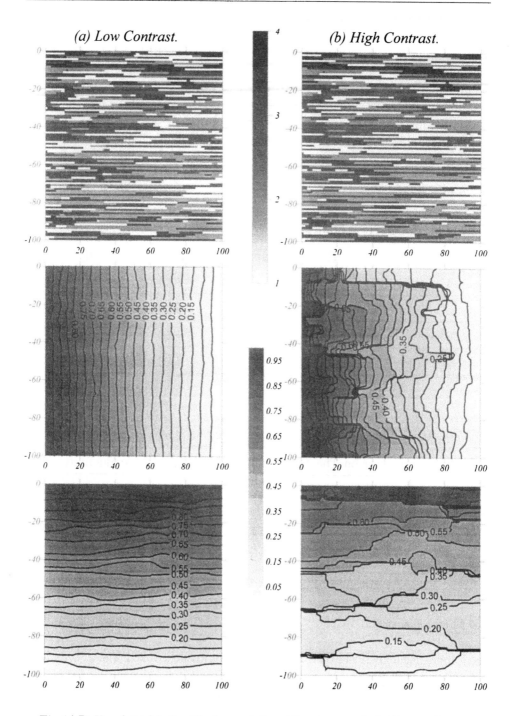

Fig.(4.7) Head Fields for Horizontal Stratifications with Different Contrasts.

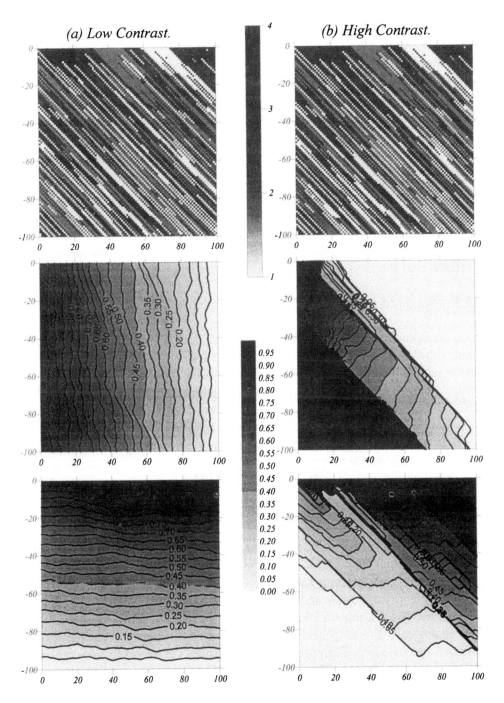

Fig.(4.8) Head Fields for Inclined Bedding with Different Contrasts.

121

Table (4.11) Flow Scenarios in The Numerical Experiments.

Scenario	Description
1	Horizontal flow in horizontal stratifications.
2	Vertical flow in horizontal stratifications.
3	Horizontal flow in inclined bedding.
4	Vertical flow in inclined bedding.

Fig.(4.7) compares the head contours in both low and high contrasts in the stratified structure under horizontal and vertical flows (scenarios 1 and 2), while Fig.(4.8) compares the head contours in both low and high contrasts in the inclined bedding structure (scenarios 3 and 4). The head pattern would look essentially the same due to the fact that the structure in both low and high contrasts is the same, but the influence of the structure in more pronounced in case of high contrast.

4.9.5 Estimation of Effective Hydraulic Conductivity

It is known from the literature that, the effective conductivity of a perfectly layered system with flow parallel to the bedding is estimated by the arithmetic mean, K_a, of the conductivities of each individual unit. In case of series arrangements of geological materials with flow normal to the system, the harmonic mean, K_h, is the effective conductivity. In case of correlated isotropic random fields, the effective conductivity is estimated by the geometric mean, K_g. The following power average formula is used which is considered as a general case [Journel et al., 1986],

$$K_{eff} = \left[\frac{1}{N} \sum_{i=1}^{N} K_i^p \right]^{1/p} \tag{4.27}$$

where, K_{eff} is the effective hydraulic conductivity, K_i is the hydraulic conductivity of the individual unit i, $i = 1,2,3.......N$. The exponent p used in averaging ranges from -1, for harmonic mean to 1 for the arithmetic mean. For $p = 0$ the formula is not valid but its limit exists and gives the geometric mean. The arithmetic and harmonic means are known as the bounds of the effective conductivities.

The geological patterns generated in this study are different from the afore mentioned special cases. Therefore, these bounds are guidelines to the expected values from the experiments. The method that will be used to estimate the effective conductivity in the present study is a common one and can be applied to all cases. The method works as follows. In case of horizontal flow from left to right, the effective conductivity K_{xxeff} is,

$$K_{xx_{eff}} = \frac{Q_x \, L_x}{L_y \, (\Phi_o - \Phi_{Lx})} \qquad (4.28)$$

where, L_x and L_y are the domain dimensions in X and Y directions respectively, Φ_o and Φ_{Lx} are the right and left boundary heads respectively, and Q_x is the seepage discharge per unit width in X direction estimated by,

$$Q_x = - \sum_{j=1}^{N_y} \left[\frac{K_{xx_{k,j}} + K_{xx_{k+1,j}}}{2} \right] \left[\frac{\Phi_{k+1,j} - \Phi_{k,j}}{\Delta x} \right] \Delta y \qquad (4.29)$$

where, k is indicator used to define the section at which the horizontal seepage flow is computed. It could be any section between 1 and N_x. A section at the middle of the aquifer has been used with $k = N_x/2$. Similarly, for vertical flow from top to bottom, formula Eq.(4.28) reads,

$$K_{yy_{eff}} = \frac{Q_y \, L_y}{L_x \, (\Phi_o - \Phi_{Ly})} \qquad (4.30)$$

where, $K_{yy_{eff}}$ is the effective hydraulic conductivity in Y direction, Φ_o and Φ_{Ly} are the top and bottom boundary heads respectively, and Q_y is the corresponding seepage discharge per unit width in Y direction estimated by,

$$Q_y = - \sum_{i=1}^{N_x} \left[\frac{K_{yy_{i,m}} + K_{yy_{i,m+1}}}{2} \right] \left[\frac{\Phi_{i,m+1} - \Phi_{i,m}}{\Delta y} \right] \Delta x \qquad (4.31)$$

where, m is indicator defining the section at which the vertical seepage flow is computed. It could be any section between 1 and N_y. A section at the middle of the aquifer has been used with $m = N_y/2$.

4.10 Results of The Experiments

4.10.1 Probability Density Function of Effective Conductivity

The numerical experiments are used to infer the probability density function of effective conductivity. The results of case 1 (low contrast) and case 2 (high contrast) are displayed in Fig.(4.9) and Fig.(4.10) respectively. The figures show the calculated probability density function of the effective conductivity which is denoted by square markers, and the theoretical normal density function is given by,

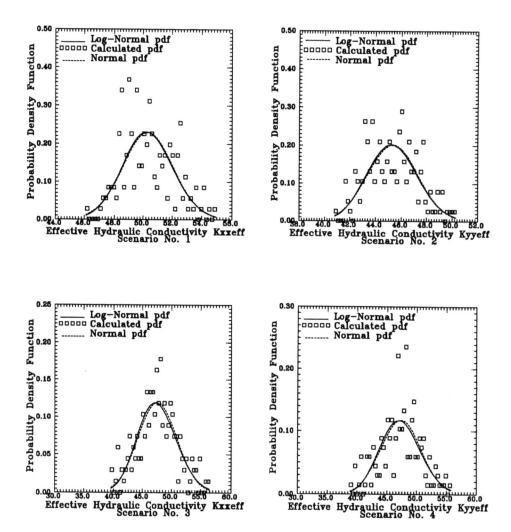

Fig.(4.9) Results of PDF form Numerical Experiments with Low Contrast.

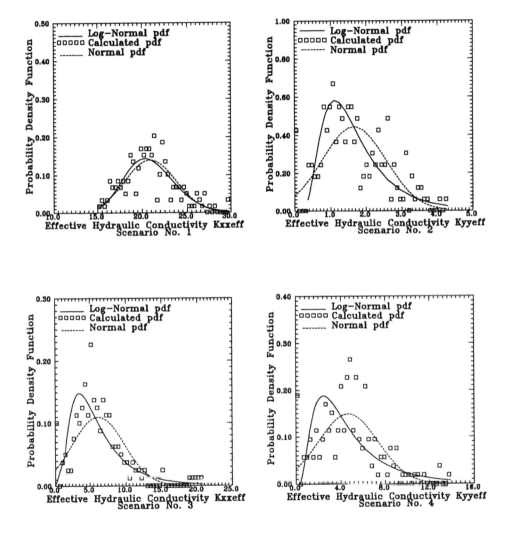

Fig.(4.10) Results of PDF form Numerical Experiments with High Contrast.

125

$$f(K_{eff}) = \frac{1}{\sigma_{K_{eff}}\sqrt{2\pi}} \exp\left[\frac{-(K_{eff} - \langle K_{eff}\rangle)^2}{2\,\sigma^2_{K_{eff}}}\right] \qquad (4.32)$$

where, $\langle K_{eff}\rangle$ is the ensemble effective conductivity and σ_{Keff} is the standard deviation of the effective conductivity. Eq.(4.32) is denoted by dotted lines, and the theoretical lognormal density function is given by,

$$f(K_{eff}) = \frac{1}{K_{eff}\,\sigma_{Ln(K_{eff})}\sqrt{2\pi}} \exp\left[\frac{-(Ln\,K_{eff} - \langle Ln\,K_{eff}\rangle)^2}{2\,\sigma^2_{Ln(K_{eff})}}\right] \qquad (4.33)$$

where, $Ln\,K_{eff}$ is the ensemble of Ln-effective conductivity and $\sigma_{Ln(Keff)}$ is the standard deviation of Ln-effective conductivity.

Eq.(4.33) is denoted by solid line in Fig.(4.9) and Fig.(4.10). Table (4.12) summarizes the ensemble statistics over the number of Monte-Carlo simulations (200 runs each) for both case studies. These estimates are used to plot the *PDF* of normal and lognormal densities in Fig.(4.9) and Fig.(4.10).

It is obvious from Table (4.12) that the coefficient of variation, *cv*, of effective conductivity in case 2 (high contrast) is more that the coefficient of variation in case 1 (low contrast). This, of course, reflects the higher degree of variability in case 2 than in case 1.

From Fig.(4.9) One can notice that there is no significance difference between fitting normal and lognormal density function to the experimental points. However, the results of case 2 (high contrast), which are displayed in Fig.(4.10) show that the lognormality is fitting the experimental points better for scenarios 2, 3 and 4. In these scenarios, the experimental points show sharper peaks which may be better represented by lognormal than normal density.

Table(4.12) Statistical Ensemble Measures of Effective Conductivity (m/day).

Case Study	No. 1 (Low Contrast)			No. 2 (High Contrast)		
Scenario	$\langle K_{eff}\rangle$	σ_{Keff}	*cv*	$\langle K_{eff}\rangle$	σ_{Keff}	*cv*
1	50.33	1.74	0.03	20.89	2.86	0.14
2	45.33	1.98	0.04	1.63	0.94	0.58
3	47.55	3.35	0.07	6.11	3.67	0.60
4	47.33	3.39	0.07	4.47	2.95	0.66

4.10.2 Check The Results with The Bounds (K_a, K_h)

The spatial statistics of a single realization for both case studies and different types of averages are displayed in Table (4.13). The mathematical expressions used for estimating the averages are given below.

(i) Arithmetic Mean K_a :

$$K_a = \sum_{i=1}^{nst} f_i \cdot K_i \qquad (4.34)$$

where, f_i = relative frequency of K_i, and *nst* is the number of the geological materials that constitute the geological system, in the present study $nst = 4$.

(ii) Standard Deviation σ_K :

$$\sigma_K = \sqrt{\sum_{i=1}^{nst} f_i \cdot K_i^2 - K_a^2} \qquad (4.35)$$

(iii) Harmonic Mean K_h :

$$K_h = \frac{1}{\displaystyle\sum_{i=1}^{nst} \frac{f_i}{K_i}} \qquad (4.36)$$

(iv) Geometric Mean K_g :

$$K_g = \prod_{i=1}^{nst} K_i^{f_i} \qquad (4.37)$$

Comparison between ensemble statistics Table (4.12) and spatial statistics of single realization Table (4.13) shows that the value of $\langle K_{xxeff} \rangle$ (50.33 m/day in case 1 and 20.89 m/day in case 2) is less than the spatial arithmetic mean K_a (50.4 m/day in case 1 and 26.80 m/day in case 2) which is expected because this case is close to the case when the flow is parallel to perfectly layered medium where the effective hydraulic conductivity is the arithmetic mean. The value of $\langle K_{yyeff} \rangle$ (45.33 m/day in case 1 and 1.63 m/day in case 2) is more than the spatial harmonic mean K_h (37.35 m/day in case 1 and 0.03 m/day in case 2) which is expected as well since this case is close to the case when the flow is normal to a perfectly layered medium. This means that the results of the experiments fall within the expected bounds.

Table(4.13) Spatial Statistics of The Single Realizations Shown in Fig.(4.6a&b).

Case Study Statistics	No. 1 (Low Contrast)		No. 2 (High Contrast)	
	Stratified Formation	Inclined Bedding	Stratified Formation	Inclined Bedding
K_a	50.4 m/day	50.0 m/day	26.8 m/day	26.2 m/day
σ_K	22.8 m/day	23.5 m/day	40.9 m/day	40.8 m/day
K_h	37.4 m/day	35.4 m/day	0.03 m/day	0.03 m/day
K_g	43.6 m/day	41.6 m/day	1.6 m/day	1.2 m/day

4.10.3 Anisotropy of Effective Conductivity

The hydraulic anisotropy of the effective conductivity is demonstrated by the present model. The anisotropy of the formation structure (structural anisotropy) is reflected in its hydraulic characteristics since ($K_{xxeff} \neq K_{yyeff}$) in case of horizontal stratified formations (scenarios 1 and 2) as shown in Table (4.12). An highly anisotropic medium appears more in case 2 than in case 1 because of the larger contrast in conductivity values.

4.10.4 Tensorial Property of The Effective Conductivity

From the present model, an approximate prove of the tensorial property could be performed. It is found that the transformation of the effective conductivities in scenarios 1 and 2 by a rotation matrix of 45° is in acceptable agreement with the values of the effective conductivities from scenarios 3 and 4. The result is achieved through the application of the following rotational transformation,

$$\begin{Bmatrix} \cos\theta & -\sin\theta \\ \sin\theta & \cos\theta \end{Bmatrix} \begin{Bmatrix} \langle K_{xxeff}^{(1)} \rangle & 0 \\ 0 & \langle K_{yyeff}^{(2)} \rangle \end{Bmatrix} \begin{Bmatrix} \cos\theta & \sin\theta \\ -\sin\theta & \cos\theta \end{Bmatrix} \cong \begin{Bmatrix} \langle K_{xxeff}^{(3)} \rangle & \langle K_{xyeff} \rangle \\ \langle K_{xyeff} \rangle & \langle K_{yyeff}^{(4)} \rangle \end{Bmatrix} (4.38)$$

where, θ is the rotational angle which is 45° in the present study, $K_{xxeff}^{(1)}$ is the effective hydraulic conductivity from scenarios 1, $K_{yyeff}^{(2)}$ is the effective hydraulic conductivity from scenarios 2, $K_{xxeff}^{(3)}$ is the effective hydraulic conductivity from scenarios 3, $K_{yyeff}^{(4)}$ is the effective hydraulic conductivity from scenarios 4, and $\langle K_{xyeff} \rangle$ is the cross term.

Comparison between the results of the right hand side and left hand side of this transformation is shown in Table (4.14). In case 1 (low contrast) the results show fairly good agreement. However, in case 2 (high contrast) the differences

128

are relatively high. The discrepancies between the values could be accounted for by the structure generation of the top boundary (both realizations have the same top boundary) which contains low permeable zones at the top corners (see Fig.(4.8)). This reduces the vertical flow and does not influence the horizontal flow so much. The left hand side values falls in the upper limits of 95% confidence intervals of the right hand sides.

Table(4.14) Checking The Validity of Eq.(4.38).

Case Study	Left Hand Side Term	Right Hand Side Term +95% conf. intervals
No. 1 Low Contrast	47.83 m/day 47.83 m/day	47.55 ± 6.68 m/day 47.33 ± 6.78 m/day
No. 2 High Contrast	11.26 m/day 11.26 m/day	6.11 ± 7.34 m/day 4.47 ± 5.90 m/day

4.11 Summary and Conclusions

A numerical simulation of groundwater flow in artificially generated heterogeneous geological formations has been elaborated. The heterogeneous structure of the geological patterns has been generated using soft information rather than hard data through the application of the coupled chain Markov model developed in Chapter 3. The resulting conductivity fields are Markovian rather than Gaussian. Solution of the governing equations is achieved through the application of a finite difference approach to the partial differential equations. The result in this research is used for two-dimensional flow problems. It can be easily extended to three dimensional problems. A Monte-Carlo approach has been followed to perform some numerical experiments to estimate the effective conductivity. The geological structure used in the experiments is in the form of horizontal stratifications and inclined bedding with long extensions and relatively small thicknesses. This sort of geological patterns appear frequently in nature, in case of the sedimentary origin of natural geological formations in fluvial environment.

The following conclusions can be drawn from this study:

(1) The present model shows quite acceptable results for flow simulations in terms of potentials, stream functions, and Darcy's velocities in highly heterogeneous formations.

(2) The results of the numerical experiments proved that there is no significant difference between fitting normal or lognormal density function to the probability

density function of the effective conductivity in case of low contrast. However, for high contrasts, the lognormal density function fits the data better. This result is in agreement with the general belief that conductivity of deposits follows a lognormal distribution.

(3) The stratifications are the main cause of hydraulic anisotropy since $\langle K_{xxeff} \rangle \neq \langle K_{yyeff} \rangle$ (scenarios 1 and 2) as shown in Table (4.12). The magnitude of the anisotropy is also related to the contrast in local conductivity.

(4) The tensorial property of the effective hydraulic conductivity could be proven, but not rigorously, from the present model. The values of the effective hydraulic conductivities estimated from scenarios 3 and 4 (horizontal and vertical flows in inclined bedding with 45°) show reasonable agreement with the corresponding values from scenarios 1 and 2 (horizontal and vertical flows in horizontal stratifications) after transformation over 45°.

CHAPTER 5

A Hybrid Stochastic Model Combining Soft Information and Hard Data

5.1 Introduction

Long time scale predictions by flow and transport models rely strongly on the appropriate description of the spatial structures of subsurface formations at various scales. These predictions are of utmost importance in connection with groundwater resource management. In the field of petroleum engineering, the geological description is also of great importance for enhancing oil recovery. Detailed knowledge of the subsurface variability is never available in a deterministic sense. An alternative is stochastic modelling. A problem in stochastic methods is that the actual formation is often modelled by unimodal stationary random fields and, as a consequence, much of the actual spatial structure is lost. Moreover, stationary fields are unrealistic in many geological settings. In practice, the natural variability observed in geological formations is nested, i.e., a geometric structure, which can be interpreted as a trend, is most often manifested with other internal fluctuations, which can be viewed as a noise, superimposed over the structure [e.g. Journel, et al., 1978; Burrough, 1983 a, b]. This practical situation, which is called multivariate multimodal nonstationary fields, is not addressed in the geohydrological literature.

From practical and economical point of view, it may not be feasible to fully characterize formation heterogeneity using hard data. Some authors [Gelhar, 1986; Journel, 1986; Freeze, et al., 1987] have suggested to incorporate subjective or soft data. This study is an attempt to integrate subjective geological information with measurements of hydraulic parameters (hard data). The geometric shapes of the geological features are modelled stochastically based on soft information using the method proposed in Chapter 3, while hard data are used to model the variability inside these geometric shapes. This will lead to an improved characterization of the subsurface heterogeneity.

5.2 Background of The Hybrid Stochastic Approach

Petroleum reservoir engineers have dealt already with the heterogeneity of oil and gas reservoirs [Halderson, et al. 1990] (some selected methods are reviewed in Chapter 2). They consider the architecture of the reservoir as the variability on megascopic scale. This scale of heterogeneity is different from the small scale variability (macroscopic scale). At the megascale one is interested in the dimensions, orientations and spatial disposition of the formation blocks (facies) of the reservoir. At the macroscopic scale, one is interested in the continuous variability inside this formation block.

Petroleum geologists use stochastic models to generate realizations of equally likely geological units within the underground reservoir or what is called facies models. The generated realizations are interpreted as characterizing the megascopic scale mentioned above. Whereas, hydrogeologists use stochastic models to generate equally likely conductivity fields. The generated realizations are viewed as describing the macroscopic scale mentioned earlier. A marriage of these two approaches is thought to be successful. Such an approach is proposed in the present chapter. The approach is referred to as the *hybrid stochastic approach*. The mixture of populations model, discussed in the next section, is the tool to describe this approach in a mathematical form.

The essence of the hybrid approach is that soft information is used to describe the architecture of the formation, reservoir discontinuities and zonal boundaries with average hydraulic properties, while hard data are used to model continuous spatial variability within each lithology. Each of these units would be treated as a separate, statistically homogeneous unit. A similar assumption was used by Brannan, et. al [1993]. In their model they assume that the geological formation is given and they only generate conductivity fields inside each unit. They also introduce lenses with different shapes and orientations in the generated fields.

5.3 Mixture of Population Model

The mixture of population models has been applied in many branches of science, e.g, pattern recognition, geology and mining. For a comprehensive survey on methodologies and applications reference is made to the literature [Titterington, et al., 1985; Everitt and Hand, 1981].

The core of the model is the analysis of field samples from different populations. Each population is characterized by its own statistical distribution and parameters. These populations are linked together with mixing weights indicating the proportion of each component in the mixture.

The mixture of populations model is applied in the present study as a two-stage approach. Firstly, the rock type differentiates each population and secondly, each individual population is modelled by multi-variate-distributions with its own mean, variance, autocorrelation function, and its statistical anisotropy pattern. This approach is different from the approaches based on fractal geometry [Hewett, 1985].

5.3.1 Mixture Probability Density Function (MPDF)

The hydraulic conductivity is considered to be log-normally distributed [Smith, 1981] with arithmetic mean $\langle K \rangle$ and variance σ^2_K. Therefore, the logarithmic transform $Y = log(K)$ is normally distributed with mean $\langle Y \rangle$ and variance σ^2_x. The values of $\langle Y \rangle$ and σ^2_Y are related to $\langle K \rangle$ and σ^2_K through the logarithmic transformation [Matalas, 1967],

$$\langle Y \rangle = \log \langle K \rangle - 0.5 \ \sigma_Y^2$$

$$\sigma_Y^2 = \log \left(1 + \frac{\sigma_K^2}{\langle K \rangle^2} \right) \tag{5.1}$$

The harmonic K_h and geometric mean K_g of a lognormally distributed K are given by,

$$K_h = \langle K \rangle \ e^{-\sigma_Y^2}$$

$$K_g = \sqrt{\langle K \rangle \ K_h} \tag{5.2}$$

Some field observations show multi-modality in the hydraulic conductivity [Jussel, et al., 1994]. This raises the idea of using a mixture of populations model. Therefore, it is reasonable to assume that the mixture probability density function of log-hydraulic conductivity, Y, is linear combinations of n normal density functions that satisfy the relation:

$$f(Y) = \sum_{i=1}^{n} w_i \cdot \frac{1}{\sigma_{Y_i} \sqrt{2\pi}} \exp\left[-\frac{(Y - \langle Y_i \rangle)}{2 \ \sigma_{Y_i}} \right] \tag{5.3}$$

where, $f(Y)$ is the mixture density function, and w_i are the mixing weights which are considered as the marginal probabilities (volume fractions) of the states in the system. Here, w_i fulfils the condition,

$$w_i > 0,$$

$$\sum_{i=1}^{n} w_i = 1 \qquad (5.4)$$

From a *PDF* point of view the log-conductivity can be classified into four kinds as shown in Fig.(5.1). The first kind, shown in Fig.(5.1-a), represents an ideally homogeneous formation described by a Dirac delta function. The second kind, Fig.(5.1-b), shows a unimodal Gaussian distribution of conductivity which represents a formation which can be described by stationary random space function. The third kind displays a formation subdivided into four ideally homogeneous units producing a multiple of delta functions as shown in Fig.(5.1-c). The fourth kind shown in Fig.(5.1-d) represents a formation consists of four heterogeneous units characterized by multi-modal Gaussian distribution. Type (d) corresponds to more realistic field situations. In Fig.(5.1d) each bell shape represents a density function of each unit and the solid line with dots represents the mixture *PDF*. If the four units have the same variance then the field is called homoscedastic, while, if the variance varies from unit to unit, the field is called heteroscedastic.

5.3.2 Statistical Moments of The Mixture PDF

The first order moment (mean) of the mixture can be obtained by,

$$\langle K_m \rangle = \sum_{i=1}^{n} w_i \langle K_i \rangle \qquad (5.5)$$

where, $\langle K_m \rangle$ is the arithmetic mean of the mixture, $\langle K_i \rangle$ is the mean of individual population i, and w_i is the mixing weight.

The second order moment (variance) can be obtained by,

$$\sigma_m^2 = \sum_{i=1}^{n} w_i \left[\sigma_{K_i}^2 + \langle K_i^2 \rangle \right] - \left[\sum_{i=1}^{n} w_i \langle K_i \rangle \right]^2 \qquad (5.6)$$

where, σ_m^2 is the variance of the mixture, and σ_{Ki}^2 is the variance of individual population i.

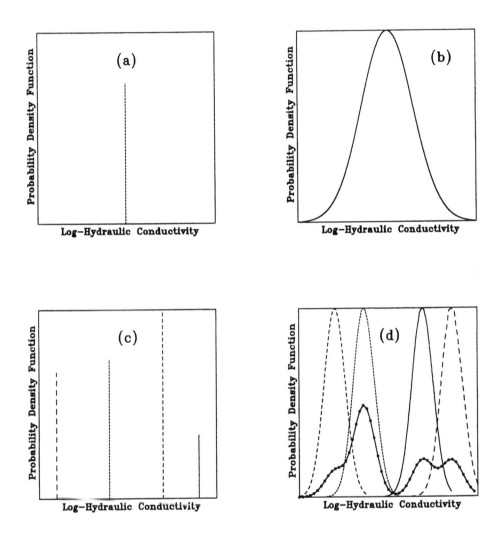

Fig.(5.1) Classification of Log-Conductivity from PDF Point of View.

5.4 Geometrical Modelling of Large Scale Variability

The purpose of this model is to identify geological zonation which describe features with discrete nature. The coupled Markov model is used to characterize this scale of variability using soft information. The subjective geological information is coded in terms of transition probabilities to feed the computer programme 'GEOSIM1'. For illustration of the model reference is made to Chapter 3.

5.5 Geostatistical Modelling of Local Scale Variability

A variety of techniques is available to generate stationary random fields with a prescribed *PDF* and an auto-correlation structure. Each of these techniques has its own advantages, disadvantages and limitations (see Chapter 2). The turning bands method is convenient in this study because it is efficient in case of a large number of grid cells. It has, with regard to other methods, also the advantage of generating ergodic fields [Tompson, et at., 1989]. The ergodicity of a field can be used to reproduce the ensemble statistics from a single realization. The algorithm of implementation is presented briefly in Chapter 2.

The log-conductivity field inside each unit is assumed to be independent from any other unit and it is modelled as

$$Y(x) = \langle Y(x) \rangle + \grave{Y}(x) \tag{5.7}$$

where, $x=(x,y)$ is a position inside a lithological unit, $Y(x)$ is log-hydraulic conductivity at position x, $\langle Y(x) \rangle$ is the expected value of log-conductivity at position x and $\grave{Y}(x)$ is the stationary Gaussian field with an expected value equal to zero and a variance equal to $\sigma^2_Y(x)$.

The field of $Y(x) = log[K(x)]$ in the various units is transformed into a field of $K(x)$ by taking the anti-logarithm of each generated value,

$$K(x) = \exp[Y(x)] \tag{5.8}$$

The spatial structure of the log-conductivity field (Gaussian field) in each geological unit is specified by an anisotropic exponential decaying autocovariance function where the principal axes of the correlation lengths are assumed to coincide with the coordinate axes. This function represents the magnitude of correlations as a function of the separation distance and vanishes when there is no correlation. It is given by,

$$C(\xi) = \sigma_Y^2 \exp\left[-\left\{\left(\frac{\xi_x}{\lambda_x}\right)^2 + \left(\frac{\xi_y}{\lambda_y}\right)^2\right\}^{1/2}\right] \tag{5.9}$$

where, ξ is the spatial lag vector, and ξ_x, ξ_y are components of the spatial lag vector ξ in x and y direction respectivity given by,

$$\xi_x = \mid x_i - x_j \mid$$
$$\xi_y = \mid y_i - y_j \mid \tag{5.10}$$

and λ_x, λ_y are the correlation lengths in x and y direction respectivity.

In geostatistics, the semi-variogram, $\gamma(\xi)$, is used instead of the auto-covariance, $C(\xi)$, [Journel and Huijbregts, 1978]. These two functions are related, under stationarity assumption, by

$$\gamma(\xi) = \sigma_Y^2 - C(\xi) \tag{5.11}$$

5.6 Implementation for Generation of Multiple Scale Field

The hybrid model is implemented in two stages. The first stage is the generation of the geological structure which corresponds to the megascale variability. The second stage is the generation of macroscale variability. These stages are elaborated as follows.

5.6.1 Generation of The Geological Structure

The geological structure is sampled over intervals of a coarse grid. Transition probabilities between the different lithologies are determined and used for the simulation. A brief description of the steps are given below (for more details and formulas reference is made to section 3.7).

(*i*) The left and top boundaries are generated by the horizontal and vertical chains respectively in the same way as described in Chapter 3 (section 3.7).

(*ii*) At each cell location in the domain, the conditional probabilities of the coupled chains are calculated for all the states.

(*iii*) A cumulative distribution of transition probabilities from the coupled chains is build.

(*iv*) Monte-Carlo sampling to realize the states of the cell given the states of the neighbours is applied.

(*v*) From first step *ii* the procedure is repeated until all cells of the simulation grid are assigned a state.

5.6.2 Generation of Macroscopic Heterogeneity

A finer grid is superimposed over the geological structure and each cell in the grid is coded with the prescribed lithology. A search algorithm is followed for assignment of values of hydraulic conductivity to each geological unit in the system. The following steps are considered.

(*i*) If x is positioned in lithology S_k, then $Y(x) = \langle Y_k \rangle + \grave{Y}_k$, $k = 1,..., n$, where $\grave{Y}_k$ is a statistically homogeneous Gaussian random field, generated by the turning bands algorithm.

(*ii*) The search algorithm is repeated to cover all the cells in the domain and each cell is assigned a value corresponds to parameters of its statistical distribution and its statistical anisotropy pattern.

(*iii*) Then the conductivity value is calculated by the anti-logarithm as, $K(x) = exp[Y(x)]$.

5.7 Illustration of The Model Results by Some Synthetic Examples

A set of non-stationary two-dimensional fields is generated by the computer models 'GEOSIM1' and 'MARKOVTB'. These fields are called here compound random fields (Markovian-Gaussian fields). The computer model can be used either to generate vertical cross-sections of formation heterogeneity at two scales of variability or a plan distribution of transmisivities with distinct zonation.

The model is illustrated by Fig.(5.2). Here, a 2D vertical cross-section of large scale geological structure is shown with dimensions of 2000 m in length and 400 m in depth (see Fig.(5.2)). It is assumed that the geological system consists of four lithological units coded 1, 2, 3 and 4 in the figure and indicated by colour white, light grey, dark grey and black respectively. The structure is generated by the Markov method described earlier with transition probabilities given in Table (5.1), sampled over intervals of 20 m in horizontal direction and 10 m in vertical direction.

A finer grid with cell dimensions of 10×5 m is superimposed over the structure for the generation of conductivity field. The conductivity variations inside each geological unit is generated by the computer model 'MARKOVTB' according to the various statistics in order to investigate various cases of non-stationary fields. Parameters in Table (5.2) are used to investigate the influence of non-stationarity in the mean, parameters in Table (5.3) are used to investigate the influence of non-stationarity in the variance, parameters in Table (5.4) are used to investigate the influence of non-stationarity in the autocorrelation pattern, and parameters in Table (5.5) are used to investigate the influence of global non-stationarity. The simulation results of these cases will be discussed in the following sections.

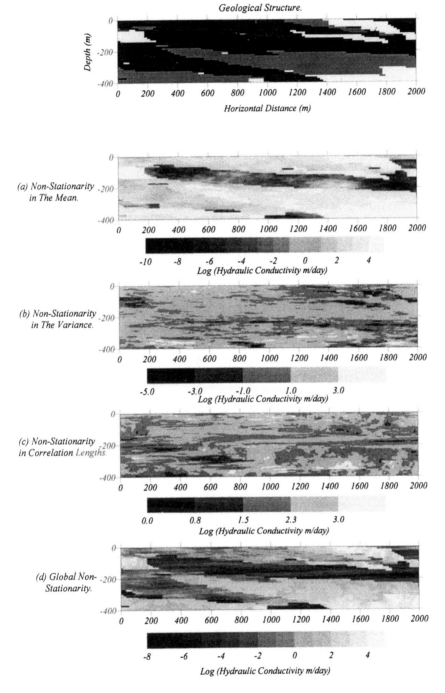

Fig.(5.2) Demonstration of The Hybrid Model with Some Synthetic Cases.

Table (5.1) Statistics of Fig.(5.2).

Length of The Given Section (m)= 2000.
Depth of The Given Section (m)= 400.
Sampling interval in X-axis (m) = 20.
Sampling interval in Y-axis (m) = 10.
No. of States = 4

Input Statistics | Calculated Statistics

Horizontal Transition Probability Matrix

State	1	2	3	4	State	1	2	3	4
1	0.960	0.010	0.010	0.020	1	0.933	0.007	0.021	0.039
2	0.010	0.970	0.010	0.010	2	0.015	0.964	0.012	0.009
3	0.010	0.020	0.960	0.010	3	0.006	0.026	0.957	0.011
4	0.010	0.010	0.010	0.970	4	0.016	0.002	0.014	0.968

Vertical Transition Probability Matrix

State	1	2	3	4	State	1	2	3	4
1	0.930	0.010	0.030	0.030	1	0.810	0.041	0.053	0.096
2	0.100	0.800	0.050	0.050	2	0.044	0.876	0.071	0.010
3	0.050	0.050	0.800	0.100	3	0.042	0.068	0.805	0.086
4	0.020	0.050	0.040	0.870	4	0.019	0.077	0.058	0.846

5.8 Discussion of Model Results

5.8.1 Case (a): Non-Stationarity in The Mean

It is assumed that each geological unit possesses a different mean conductivity, while the variance and correlation lengths are assumed equal. The simulation parameters used are displayed in Table (5.2). The theoretical and realized *PDFs* of individual units and the mixture are presented in Fig.(5.3). The *PDF* of the mixture shows multi-modality. This type of density has been reported for field measurements on gravel outcrop [Jussel, et al., 1994]. Fig.(5.2a) shows the generated conductivity field. Fig.(5.4) shows some conductivity profiles. In these figures it can be seen that the boundaries between the geological units are still clear and not smeared by the influence of the local variability, in this case due to the large contrast in the mean.

Table (5.2) Simulation Parameters (Non-Stationarity in The Mean).

Parameter State	w_i	$\langle K \rangle$ m/day	σ_K m/day	$\langle Y \rangle$	σ_Y	λ_x (m)	λ_y (m)
1	0.13	100.	10.	4.6	0.10	200.	50.
2	0.31	50.	10.	3.9	0.20	200.	50.
3	0.31	10.	10.	1.96	0.83	200.	50.
4	0.25	1.	10.	-2.3	2.15	200.	50.

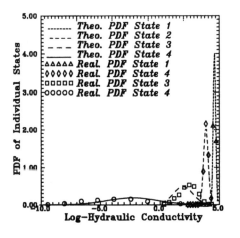 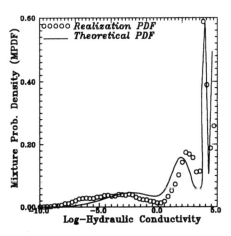

Fig.(5.3) Individual and Mixture PDF (Non-Stationarity in The Mean).

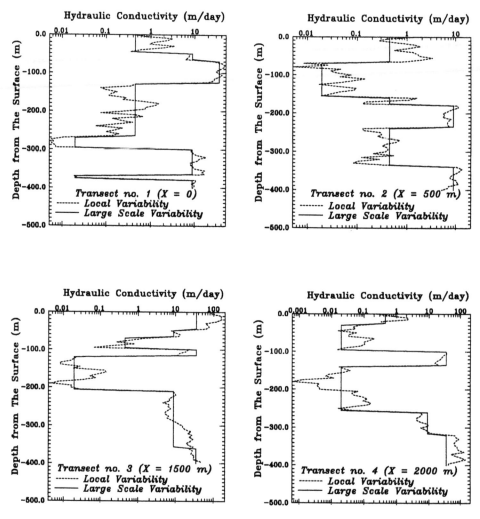

Fig.(5.4) Individual and Mixture PDF (Non-Stationarity in The Variance).

5.8.2 Case (b): Non-Stationarity in The Variance

It is assumed that the geological units possess different conductivity variances, while the arithmetic mean and the correlation lengths are constant. The simulation parameters used are displayed in Table (5.3). The theoretical and realized *PDFs* of the individual units and of the mixture are presented in Fig.(5.5). In this case one notices that the mixture *PDF* is unimodal but skewed to the right and it has an elongated tail towards the left. This indicates that high permeable zones are more likely to appear in this formation. Fig.(5.2b) shows the conductivity field and Fig.(5.6) shows some conductivity profiles. In these figures it can be seen that the geological boundaries are smeared because the units are having the same mean. Hence, the individual densities are overlapping such that a single-peaked density arises.

Table (5.3) Simulation Parameters (Non-Stationarity in The Variance).

Parameter State	w_i	$\langle K \rangle$ m/day	σ_K m/day	$\langle Y \rangle$	σ_Y	λ_x (m)	λ_y (m)
1	0.13	5.	20.	0.19	1.68	100.	10.
2	0.31	5.	10.	0.80	1.27	100.	10.
3	0.31	5.	5.	1.26	0.83	100.	10.
4	0.25	5.	2.	1.54	0.39	100.	10.

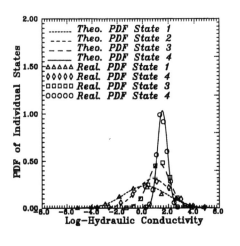 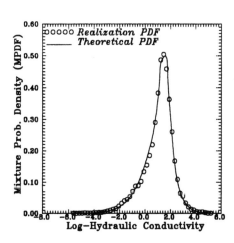

Fig.(5.5) Conductivity Profiles in Case of Non-Stationarity in The Mean.

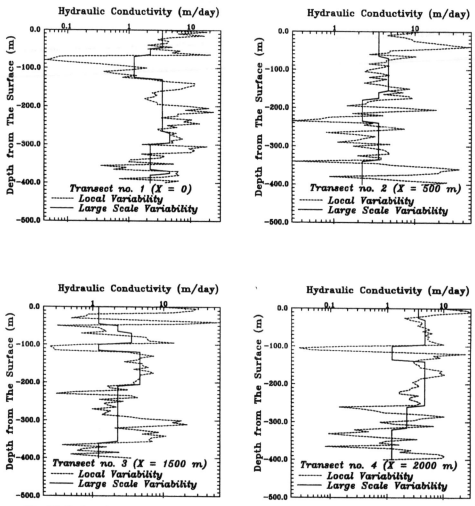

Fig.(5.6) Conductivity Profile in Case of Non-Stationarity in The Variance.

5.8.3 Case (c): Non-Stationarity in The Correlation Lengths

It is assumed that the geological units possess different correlation lengths but the arithmetic mean of conductivity, the variance and the shape of the auto-covariance function are equal. The parameters used are displayed in Table (5.4). The theoretical and realized *PDFs* of individual units and of the mixture are presented in Fig.(5.7). In this case it is essential that the mixture *PDF* is unimodal and represented by Gaussian-type distribution. One also can notice that the *PDF's* of the individual populations coincide because all units have the same statistical moments. Fig.(5.2c) shows the conductivity field and Fig.(5.8) shows some conductivity profiles. In these figures it can be seen that the boundaries are smeared, but from the local variability one can notice different correlation patterns in the generated field.

Table (5.4) Simulation Parameters (Non-Stationarity in Correlation Lengths).

Parameter State	w_i	$\langle K \rangle$ m/day	σ_K m/day	$\langle Y \rangle$	σ_Y	λ_x (m)	λ_y (m)
1	0.13	10.	5.	2.19	0.47	50.	50.
2	0.31	10.	5.	2.19	0.47	100.	50.
3	0.31	10.	5.	2.19	0.47	200.	10.
4	0.25	10.	5.	2.19	0.47	500.	5.

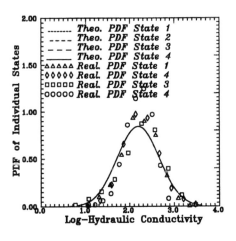
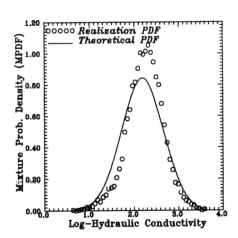

Fig.(5.7) Individual and Mixture PDF (Non-Stationarity in Corr. Lengths).

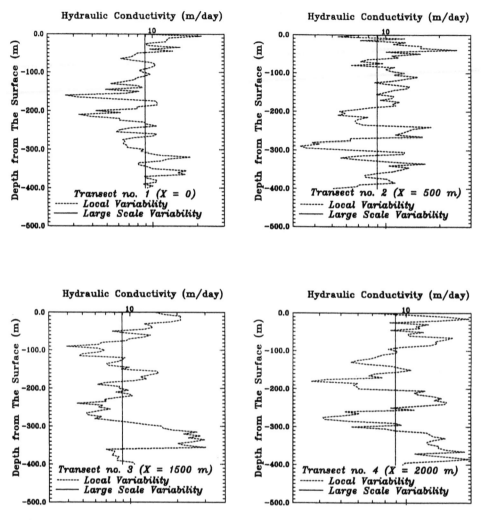

Fig.(5.8) Conductivity Profiles in Case of Non-Stationarity in Corr. Lengths.

5.8.4 Case (d): Global Non-Stationarity

It is assumed that the geological units possess different parameters: arithmetic mean of hydraulic conductivity, variance and correlation lengths, while the auto-covariance functions are equal. This case represents a real field situation. The simulation parameters used are displayed in Table (5.5). The theoretical and realized *PDFs* of individual units and of the mixture are presented in Fig.(5.9). In this case it is clear that the mixture *PDF* is multimodal, in this case characterized by two peaks. Fig.(5.2d) shows the conductivity field generated and Fig.(5.10) shows some conductivity profiles. In these figures it can be seen that the boundaries are remarkable like case (a). The non-stationarity in the mean is therefore the major reason of having pronounced boundaries.

In transect number 4 of Fig.(5.10) the local variability denoted by the dotted lines is lower than the large scale variability (mean) denoted by the solid line in unit 4. This is because the mean is representative for the whole unit. The local variability can be more than the mean in other transects such as number 1 and 3.

Table (5.5) Simulation Parameters (Global Non-Stationarity).

Parameter State	w_i	$\langle K \rangle$ m/day	σ_K m/day	$\langle Y \rangle$	σ_Y	λ_x (m)	λ_y (m)
1	0.13	50.	50.	3.57	0.83	50.	50.
2	0.31	10.	5.	2.19	0.47	100.	50.
3	0.31	1.0	2.	-0.80	1.0	200.	20.
4	0.25	0.1	0.5	-3.93	1.81	500.	5.

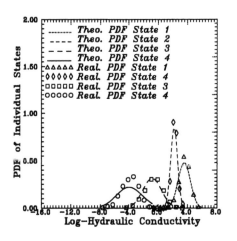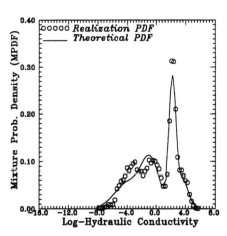

Fig.(5.9) Individual and Mixture PDF (Global Non-Stationarity).

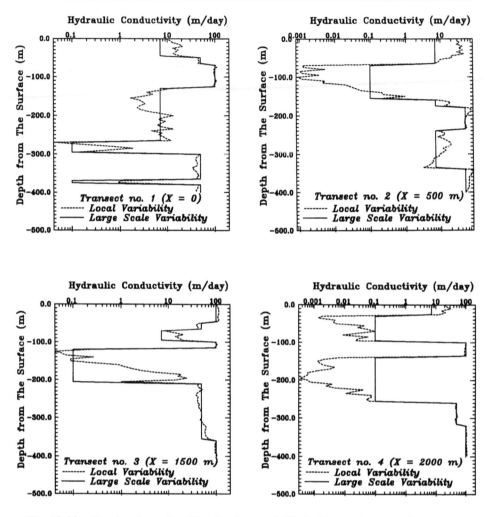

Fig.(5.10) Conductivity Profiles in Case of Global Non-Stationarity.

5.8.5 Interpretation of The Model Results by Variogram Analysis

Variograms of *log* (*K*) have been estimated in *x* and *y* directions for all previous cases in order to interpret the influence of various types of non-stationarity on the global variogram behaviour. This may help in analyzing variograms of real field situation. Fig.(5.12) and Fig.(5.13) display variograms of all cases mentioned earlier, in *x* and *y* directions respectively.

For all the cases mentioned above, equivalent globally stationary fields are generated to evaluate the validity of the global-stationarity assumption. The fields are presented in Fig.(5.11). The variograms of the fields are calculated and plotted in the same figures (Fig.(5.12) and Fig.(5.13)) for comparisons. Table (5.6) displays the statistical parameters used to generate the equivalent stationary fields. These parameters are estimated from the mixture moments by Eq.(5.5) and Eq.(5.6).

In case (a), non-stationarity in the mean, the non-stationarity in the variogram is most pronounced compared to the other cases, since the variogram does not seem to reach a constant value (sill) and it is growing with the spatial lag. From Fig.(5.12a) and Fig.(5.13b) it is clear that the stationary variogram is far from the non-stationary variogram, which means that the assumption of globally stationarity field in this case is not applicable.

In case (b), non-stationarity in the variance, the non-stationarity behaviour is less pronounced. The reason for that is the influence of high variances and constant mean which smear the boundaries between the different geological units. This can be observed clearly in the profiles of Fig.(5.6). Fig.(5.12b) and Fig.(5.13b) show a reasonable agreement (practically) between variogram shapes in case of non-stationarity and globally stationary field. One can notice that the use of an equivalent globally stationary field to represent the non-stationary is not proper although the variogram shapes look relatively close at further lags. This means that sometimes non-stationarity of the data cannot be observed clearly in the variogram.

In case (c), non-stationarity in the correlation lengths, the variogram looks quite stationary. This means that there is no significant influence of non-stationarity of the correlation patterns of the individual units on the global variogram behaviour. Estimation of the (asymptotic) sill variance from Fig.(5.12c), which is about 0.22 is in perfect agreement with the variance from Table (5.4) that is $(0.47)^2 = 0.221$. This ensures good reproduction of the variance in the simulated field. Fig.(5.12c) and Fig.(5.13c) show a relatively fair agreement in the variorgam shapes in case of the non-stationary field and its equivalent globally stationary field. This indicates that non-stationarity in correlation patterns can not be observed in the variogram.

Table (5.6) Simulation Parameters for Equivalent Global Stationary Fields.

Parameter Case	$\langle K_m \rangle$ m/day	σ_m m/day	$\langle Y_m \rangle$	σ_{mY}	$\langle \lambda_x \rangle$ (m)	$\langle \lambda_y \rangle$ (m)
a	31.85	34.5	3.07	0.881	200.	50.
b	5.	10.8	0.74	1.32	100.	10.
c	10.	5.	2.19	0.47	224.5	26.
d	9.935	26.3	1.26	1.44	224.5	29.5

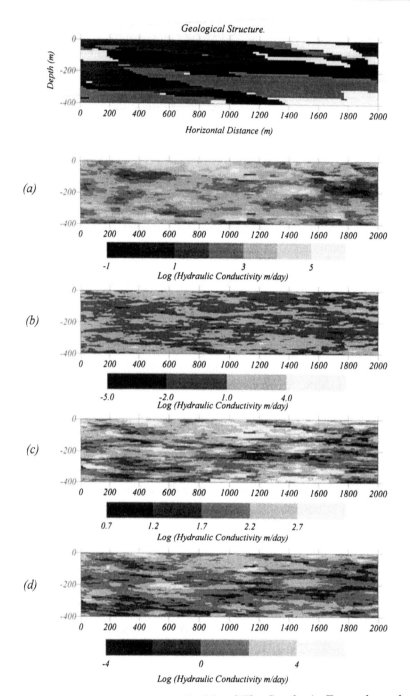

Fig.(5.11) Equivalent Stationary Fields of The Synthetic Examples schown in Fig.(5.2).

150

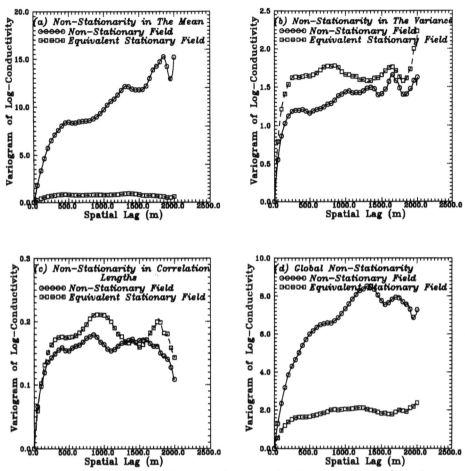

Fig.(5.12) Variorgams of Log-Conductivity for All Cases in X-direction.

In case (d), globally non-stationarity, the variograms in Fig.(5.12d) and Fig.(5.13d) are far apart, as similar to case (a). This can be attributed to the distinction in the mean conductivities of different geological units which are quite pronounced. The assumption of equivalent stationarity field does not hold, which means that serious errors can be introduced due to neglecting non-stationarity.

5.9 Procedure for Field Applications

5.9.1 Sampling Network

The application of the hybrid model requires a two-scale sampling network. A sampling network for local variability is required. In this network one or more

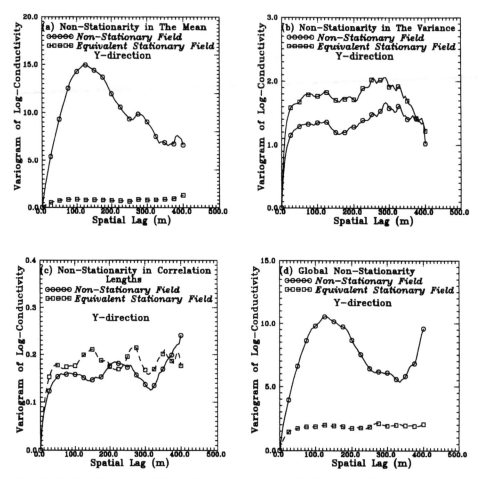

Fig.(5.13) Variograms of Log-Conductivity for All Cases in Y-direction.

clusters of closely spaced sampling stations are needed to provide estimates of parameters at the local scale (means, variances and correlation lengths). Another sampling network for megascale variability is required by means of a number of scattered stations to sample different geological units in the reservoir. These stations provide the information needed to obtain the reservoir structure and an estimation of transition probabilities.

5.9.2 Acquisition of Soft Information

Soft information can be in the form of expert experience, physical reasoning and common sense about the geological environment in which the formation is deposited, similar sedimentation patterns from better known sites, geological investigations of open pits and visual observation of depositional patterns from analogous outcrops with good geological judgment based on geological experience. This kind of soft geological information can be incorporated to construct subjective cross-sections of the spatial distribution of lithological units. This information can be translated to a hand-drawn picture of what the formation is expected to be. Then the picture is coded by assigning each geological material a code 1, 2, 3...etc. The transition probabilities are estimated according to the procedure mentioned in Chapter 3 (section 3.6.2).

5.9.3 Acquisition of Hard Data

A-priori estimates of the hard data (mean, variance, correlation lengths) are needed to perform a geostatistical analysis. Laboratory tests and/or field measurements of the local hydraulic conductivity can be performed on field samples at different locations (multi-level samplers) such as the Borden site [Woodbury, 1991]. In many cases such measurements are not available, but still the estimates can be made with some degree of subjective accuracy, e.g., these estimates can be obtained from the analyst's expertise and common sense, and available data from geologically similar sites. Some data can be obtained from the literature [Gelhar, et al., 1992; Dagan, 1989]. Conductivities may be inferred from some alternative parameters, such as grain size distribution by performing sieve or wet analysis on samples and cores.

An important but difficult task is to postulate the underlying variogram or the auto-covariance model (linear, spherical, Gaussian, exponential, etc.) and to estimate the unknown parameters (correlation lengths, nugget, etc.) in the model from the available data. Since the variogram or the autocovariance functions are the basic tools on geostatistics, Journel & Huijbregts [1978] suggest that a minimum of 30 pairs of points is necessary to estimate reliable parameters of the autocovariance function.

5.10 Summary and Conclusions

Traditional random field models are used to generate fields of hydrogeological parameters which are appealing from a statistical point of view, but they poorly represent geological features, particularly when the field contains some pronounced and known geological attributes. Therefore, stochastic models must be guided by classical geological knowledge. Incorporating soft geological information into a geostatistical simulation would lead to more plausible realizations of site specifications not only from a statistical point of view but also from a geological view point.

This chapter proposes a methodology for the incorporation of soft geological information in addition to the available hard data in order to characterize geological heterogeneity under two scales of variability. A mixture-of-populations model is used as a tool to mathematically describe this type of heterogeneity.

The following concluding remarks are drawn:

(1) Geological knowledge and expertise of subjective nature (translated into transition probabilities) can be combined with quantitative geostatistical models (in terms of correlated random fields) to produce more accurate realizations of lithology and hydraulic conductivity fields of underground reservoirs.

(2) The proposed hybrid model can be used to generate realizations of conductivity within the geological structure, with all realizations satisfying the 'soft' geological knowledge (reproduction of transition probabilities) and 'hard' data (reproduction of *PDF*s and its moments). The model can be extended to honour hard data and their locations by implementing conditioning using a Kriging algorithm mentioned in Chapter 2. This point may be useful to be considered in future research.

(3) The representation of spatial variability by this model is promising and it provides a good step for introducing multiple-scale stochastic processes. Applications of this model in the form of a single realization approach to porous flow and transport problems will be discussed in Chapter 6.

(4) The hybrid model allows the use of multiple realizations in order to evaluate uncertainty of the flow and transport predictions under at least, in this case, two scales of variability: the lithological scale (megascale variability) and the individual unit variability (macroscale variability). Applications of the model using multiple realizations will be discussed in Chapter 7.

(5) Large scale field studies on spatial variability, as suggested in section 5.9, would be a proper test to evaluate and validate the proposed hybrid model.

(6) From the synthetic examples, it has been shown that high contrast in the mean of individual populations is the main factor of pronounced non-stationarity in comparison to other types of non-statinarity.

(7) The assumption of a globally stationary random function model is not adequate to describe many fields as far as the non-stationarity in the mean is concerned.

(8) From the analysis of *PDF*s it is shown that the boundaries between the geological units are remarkably clear when the *PDF* is characterized by multiple peaks.

(9) Unless the non-stationarity of the data is clearly pronounced, it can not be manifested by variogram analysis.

(10) The uni-modal multi-variate stationary field can be considered as a special case of the multi-modal multi-variate non-stationary field which can be generated by the proposed model. This conclusion is best underlined by case (c) which is characterized by a single-peaked *PDF*.

154

CHAPTER 6

Flow and Transport in Heterogeneous Formations: Single Realization Approach

6.1 Introduction

In the literature on stochastic theory of flow and transport, spatial variability of aquifer parameters is modelled as stationary random fields (see Chapter 2). For example, stochastic theories reviewed by Dagan [1989], [Neuman, et al., 1987] and more recently by Gelhar [1993] are based on the characterization of the hydrogeological parameters as a stationary random space functions described by its mean, variance and its autocorrelation structure either in terms of its auto-covariance or variogram function. In many real geological situations this characterization is not sufficient.

Just recently, new theories of flow and transport in multiple scale random fields have been developed [e.g. Cushman & Ginn, 1993; Dagan, 1994; Glimm et al., 1993]. These authors assume that the porous medium possesses a fractal character. Fractals are characterized by the fact that they manifest similar variations at all length scales of observations (for more details see Hewett, 1985). Although this assumption has not yet been supported by field observations, a new research is directed to this concept [Wheatcraft, et al., 1988; Cushman & Ginn, 1993; Sahimi, 1995].

Most of the recently developed stochastic theories and applications do not take into account the geometrical architecture of the underground reservoir. More reliable predictions of transport in geological media require the integration of the effects from many different scales. The heterogeneity may be discrete (e.g. geometrical heterogeneity), continuous (e.g. parametric variability) or compound (geometrical and parametric variability).

The approach described in this chapter, however, is an intermediate step between the unimodal stationary Gaussian approach of the classical theories and the recently developed fractal approach.

Because of the discrete nature of the Markov model (Chapter 3), because of the difficulties which one may face, and because of the simplifying assumptions which one has to make in the derivation of analytical expressions, a numerical method has been applied. Numerical models have the advantages of being more general and flexible for addressing geological variability. To the best of the author's knowledge, there is no successful analytical approach (stochastic differential equations) that can address all nonlinearities, multi-dimensionality, global non-stationarity, and multi-facies reservoir in flow and multi-species transport modelling.

To the author's knowledge, yet no attempts are made to recognize the multiple scales of variability on transport characteristics by numerical experiments. This study focuses on the influence of discrete (Markovian), continuous (Gaussian) and compound (Markovian-Gaussian) heterogeneity on flow and transport behaviour.

6.2 Purpose of The Numerical Experiments

The purpose of the experiments is to investigate some aspects of transport processes in physically plausible heterogeneous structures and to compare results with analytical solutions for some special cases.

The final goal of transport simulation studies is, in general, prediction of concentration fields, macrodispersion coefficients, breakthrough curves or travel time distributions, etc. The effect of various geological settings and characterization methodologies on these transport parameters is the main objective of these experiments.

6.3 Design of The Experimental Programme

The experimental programme contains the realization of heterogeneous fields generated by various methods (Gaussian, Markovian and compound) and the solution of flow and transport in these fields. The various types of heterogeneity models used and the method of solving flow and transport are discussed in next sections.

6.3.1 Characterization Methodologies

6.3.1.1 Markovian Heterogeneity Model

The Markov model referred to here, is discussed in Chapter 3. Computer-programmes GEOSIM1 and GEOSIM2 are developed based on this method. Some realizations will be displayed in this chapter.

6.3.1.2 Gaussian Heterogeneity Model

The turning bands method, described in Chapter 2, has been programmed in a code called 'TBG' and used to generate a log-normally stationary two-dimensional hydraulic conductivity field with a prescribed mean, variance and exponential autocovariance function. The spectral method [Mantoglou and Wilson, 1982] was used to generate the line process. A set of 100 evenly spaced distribution lines was used. The number of spectral harmonics was 100. The statistics of the synthesized hydraulic conductivity fields were reproduced with a reasonable degree of accuracy.

6.3.1.3 Compound Heterogeneity (Markovian-Gaussian) Model

The hybrid model developed in Chapter 5 is used to generate this type of heterogeneity. The generated field possesses Markovian characteristics in terms of transition probabilities between the states (geological units) and Gaussian properties of the parametric variability inside each state. The algorithm is elaborated in Chapter 5.

6.3.2 Synthetic Realizations Used in The Experiments

The following two-dimensional vertical cross sections of hypothetical formations with characteristics similar to those of some actual media are presented:
   (1) Homogeneous field (test case).
   (2) Perfectly layered formation (test case).
   (3) Single heterogeneity (inclusion).
   (4) Imperfectly stratified formation (Markovian fields).
   (5) Stationary Gaussian random function with exponential autocorrelation (Gaussian fields).
   (6) Large scale formation with large size geological units.
   (7) Large scale formation with multiple scales of variability (compound fields).
For the aforementioned formations the following scheme is implemented:
   (*i*) High and low permeability contrast of the geological units.
   (*ii*) Effect of pore scale dispersion.
   (*iii*) Effect of two scales of variability (Megascopic and Macroscopic scales).
   (*iv*) Influence of various characterization methodologies.

6.3.3 Discretization of Hydraulic Conductivity Fields

In order to obtain statistically meaningful results from a single replicate, the stationary hydraulic conductivity field should fulfil some conditions. According to Ababou [1989], the suggested conditions can be,

$$\Delta s \leq \frac{\lambda_s}{4}, \quad s = x, y \tag{6.1}$$

where, Δs is the domain disceretization and λ_s is the correlation length, and the domain dimension is related to the correlation lengths by,

$$L_i \geq 25\lambda_i, \quad i = x, y \tag{6.2}$$

where, L_i is the domain dimension. In the experiments, the cell hydraulic conductivities are assumed isotropic. The variability in porosity is of secondary importance [Dagan, 1986], and subsequently for the porosity a value of 0.3 is assumed for all experiments.

6.4 Modelling Groundwater Flow

The hydrodynamic condition for the flow system is a steady state groundwater flow in a two-dimensional nondeformable saturated porous medium which is governed by the mass-conservation equation,

$$\nabla \cdot q = 0 \tag{6.3}$$

where, q is the local specific discharge.
Introduction of Darcy's law,

$$q = -K \nabla \Phi \tag{6.4}$$

where, K is the hydraulic conductivity tensor and Φ is the potential head. This leads to the well known elliptic *PDE*,

$$\nabla \cdot [K \nabla \Phi] = 0 \tag{6.5}$$

6.4.1 Computational Algorithm of The Flow System

The considered flow domain is rectangular, Fig.(6.1), with length L_x and width L_y. A mean flow is driven in the horizontal direction (x-direction) from left to right by imposing a head gradient $J_x = -\Delta\Phi/L_x$ obtained from a potential difference $\Delta\Phi = 1.0$ over length L_x on the field. The lower and upper boundaries are impervious. The flow equation, Eq.(6.5), is discretized using an implicit finite difference scheme with five-points operator. The resulting linear set of equations is solved using the conjugate gradient method (see Chapter 4). The computed potential head

158

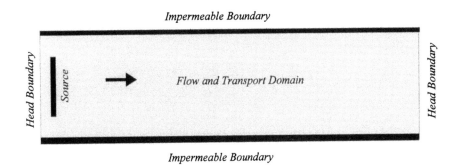

Fig.(6.1) Definition Sketch of The Flow and Transport Problem.

distribution is differentiated to derive the velocity. The details of the developed flow models 'MEGAFLOW' for potential head calculations and 'MEGASTR' for stream functions are given in Chapter 4.

6.5 Modelling Solute Transport

6.5.1 Governing Equations

This study considers a single contaminant species with high solubility and its density and viscosity are similar to that of the water such as chloride or bromide tracers. There is no interaction of the solute with the solid matrix (inert solutes). A transient plume migration in a steady velocity field is studied. The two-dimensional equation that describes solute transport in these circumstances, taking into account advection, dispersion and diffusion, can be written in the following form [Bear, 1972],

$$\frac{\partial C}{\partial t} + V_x \frac{\partial C}{\partial x} + V_y \frac{\partial C}{\partial y} - \frac{\partial}{\partial x}\left[D_{d,xx}\frac{\partial C}{\partial x} + D_{d,xy}\frac{\partial C}{\partial y}\right]$$
$$- \frac{\partial}{\partial y}\left[D_{d,yx}\frac{\partial C}{\partial x} + D_{d,yy}\frac{\partial C}{\partial y}\right] = 0 \qquad (6.6)$$

where, C is the concentration field at time t, V_x and V_y are the components of the Eulerian velocity field in x and y direction respectively defined as follows,

159

$$V_y = -\frac{K}{\varepsilon}\frac{\partial \Phi}{\partial y} \qquad (6.7)$$

$$V_x = -\frac{K}{\varepsilon}\frac{\partial \Phi}{\partial x} \qquad (6.8)$$

where, ε is the effective porosity, K is the isotropic hydraulic conductivity, $D_{d,xx}$, $D_{d,xy}$, $D_{d,yx}$ and $D_{d,yy}$ are components of pore scale (micro-level) dispersion tensor [Bear, 1961b],

$$D_{d,ij} = \left(\alpha_t |V| + D_m\right)\delta_{ij} + \left(\alpha_l - \alpha_t\right)\frac{V_i V_j}{|V|} \ , \ ij = 1,\ 2 \qquad (6.9)$$

where, δ_{ij} is the Kronecker delta, $\delta_{ij}=1$ for $i=j$ and $\delta_{ij}=0$ otherwise, α_l is the longitudinal dispersivity, α_t is the transversal dispersivity, D_m is the molecular diffusion coefficient, the indices 1 and 2 correspond to x and y direction respectively and $|V|$ is the magnitude of the resultant velocity given by,

$$|V| = \sqrt{V_x^2 + V_y^2} \qquad (6.10)$$

6.5.2 Initial and Boundary Conditions

For the initial conditions of the solute concentration, it is assumed that no contaminants are present in the system, so that $C(x,y,0)=0$, for $0 \leq x \leq L_x$, $0 \leq y \leq L_y$. The boundary condition, $\partial C/\partial y=0$ at $t>0$, $0 \leq y \leq L_y$ is imposed at the top and the bottom boundaries of the flow domain. The source is located at the upstream side with some distance apart from the left boundary as shown in Fig.(6.1).

6.5.3 Random Walk Model

The classical approach to contaminant transport problems is based on Eulerian methods such as finite difference (*FDM*), finite element method (*FEM*), Lagrangian methods like random walk (*RW*) and Eulerian-Lagrangian methods such as method of characteristics (*MOC*). The Random walk model has proven to provide good approximation of the dispersion process. It has the ability to solve problems having zero or low dispersivities (large Peclet number) and it has also the advantage of having no numerical (artificial) dispersion, it does not produce negative concentrations, conservation of mass is satisfied, and it is easy to implement. Therefore, the transport equation, Eq.(6.6), is solved using the random walk. The idea is that, the injected mass is discretized into a large number of particles. Each particle carrying a part of the injected mass is convected with a deterministic velocity calculated from the flow system, and a random displacement calculated from the pore scale local dispersivity. The use of random walk method for transport simulation is done through the analogy of Eq.(6.6) and the Fokker-Planck equation [Uffink, 1990]. The particle tracking algorithm used here is similar to the one described by Kinzelbach and Uffink [1989]. It provides an accurate way of tracking a large number of particles through a fine nodal grid discretizing a highly variable velocity field. The random walk equation of a particle in a two-dimensional flow field is written explicitly as [Kinzelbach, 1986],

$$X_p(t+\Delta t) = X_p(t) + V_x(X_p(t), Y_p(t)).\Delta t + \left(\frac{\partial D_{d,xx}}{\partial x} + \frac{\partial D_{d,xy}}{\partial y}\right)\Delta t + \quad (6.11)$$

$$\frac{V_x}{|V|} Z.\sqrt{2\alpha_l \ |V(X_p(t),Y_p(t))|.\Delta t} - \frac{V_y}{|V|} Z'.\sqrt{2\alpha_t \ |V(X_p(t),Y_p(t))|.\Delta t}$$

$$Y_p(t+\Delta t) = Y_p(t) + V_y(X_p(t), Y_p(t)).\Delta t + \left(\frac{\partial D_{d,yx}}{\partial x} + \frac{\partial D_{d,yy}}{\partial y}\right)\Delta t + \quad (6.12)$$

$$\frac{V_y}{|V|} Z\sqrt{2\alpha_l \ |V(X_p(t),Y_p(t))|.\Delta t} + \frac{V_x}{|V|} Z'\sqrt{2\alpha_t \ |V(X_p(t),Y_p(t))|.\Delta t}$$

where, $(X_p(t), Y_p(t))$ are the x and y coordinates of a particle at time t, $(X_p(t+\Delta t), Y_p(t+\Delta t))$ are the x and y coordinates of a particle at time $t+\Delta t$, Δt is the time step of calculations, Z, Z' are two independent random numbers drawn from normal distribution. This produces displacements with zero mean and a covariance $2D_{d,ij}\Delta t$.

On the right hand side of Eqs.(6.11) and (6.12), the first term is the old position of the particle, the second term is the convective displacement, the third term is the so-called Fokker-Plank term and the last two terms are the dispersive components projected in the x and y directions respectively.

6.5.4 The Computational Framework of Tracing The Particles

Two-dimensional non-reactive advective-dispersive transport of a single species is considered in a uniform and nonuniform velocity field. The direct implementation of the Lagrangian concept is to release a large number of particles (5000 are used in all the numerical experiments) which constitutes the solute mass into the flow system at time $t=0$ from a prescribed source located in the upstream of the domain (see Fig.(6.1)).

6.5.4.1 Source Release Zone

In the experiments, the solute source is represented by rectangular strip where the particles are distributed randomly and uniformly in the release zone with the longer side in a direction perpendicular to the flow direction as shown in Fig.(6.1). The initial particles' coordinates are calculated by,

$$X_p(0) = X_o + \left(0.5 - [r]_0^1\right) W_o/2$$
$$Y_p(0) = Y_o + \left(0.5 - [r]_0^1\right) D_o/2 \tag{6.13}$$

where, W_o, D_o are the dimensions of the input zone (initial solute body), $(X_p(0), Y_p(0))$ are the initial coordinates of the particles, (X_o, Y_o) are the centroid of the initial solute body at time $t=0$ and r is a uniform random number between 0 and 1.

The values of W_o=1 m and D_o=30 m are used in all the experiments except in the test cases where point source is used in homogeneous medium and a fully penetrating line source for layered medium.

All the particles carry an equal mass m_p which is given by,

$$m_p = \frac{M_o}{N_p} \tag{6.14}$$

where, M_o is the total mass of solute, and N_p is the number of particles.

A value of M_o=100 grams is used for all the experiments. The centroid of the source is located 5 m from the left edge of the domain. The reflection principle is applied at the impervious boundaries [Uffink, 1990].

6.5.4.2 Criterion for Time Step

Time step is chosen sufficiently small to make sure that particles move continuously from cell to cell and any overshoot in the field is avoided, i.e., the particle should not move more than one cell during a time step. This criterion is similar to the well-known Courant-Friedrichs condition used in finite difference schemes for the pure convection transport equation which ensures that the concentrations follow the characteristic lines of the *PDE* and numerical instability is avoided. For this reason the time step is chosen according to,

$$\Delta t = t_f \frac{\Delta x}{V_{max}} \tag{6.15}$$

where, Δx is the cell dimension in the mean flow direction (x-direction), V_{max} is the maximum velocity in the flow field and t_f is called the time factor ($t_f \le 1$) which is taken to be equal to one in the calculations.

6.6 Statistical Analysis of Particles' Displacements

The method of moments is followed to detect the behaviour of the plume spreading. This approach will serve in the analysis of the Fickian dispersion regime. The *ij*th spatial moment of a concentration field at certain time is defined according to [Freyberg, 1986],

$$ M_{ij}(t) = \int\int_{-\infty}^{\infty} x^i \, y^j \, \varepsilon \, C(x,y,t) \, dxdy \qquad (6.16) $$

Freyberg further defines, zero spatial moment:
M_{00} = mass of the solute in the solution,
first spatial moment:
$X_c = M_{10}/M_{00}$ x coordinate of the center of mass,
$Y_c = M_{01}/M_{00}$ y coordinate of the center of mass,
(X_c, Y_c) is the plume centroid,
second spatial moment:
$\sigma^2_{xx} = M_{20}/M_{00} - X_c^2$ spatial variance about X_c,
$\sigma^2_{yy} = M_{02}/M_{00} - Y_c^2$ spatial variance about Y_c,
$\sigma^2_{xy} = M_{11}/M_{00} - X_c Y_c$ spatial covariance in x-y plane, and
$\theta = .5 \, arctan[2\sigma^2_{xy}/(\sigma^2_{xx} - \sigma^2_{yy})]$ angle of rotation of the plume.
σ_{xx} is called the mixing length.
The second spatial moments can be viewed as the moments of inertia of the solute body. These moments characterize the plume spreading around its centroid. It will change as the plume moves through the heterogeneous velocity field, and its spatial geometry will vary erratically with time. During the experiments, evolution of these moments is recorded with the growth of time.

6.7 Mixing Mechanisms

The mixing mechanisms of dispersion differ from one scale to another. At the molecular scale, diffusion causes mixing. At the microscopic scale, mixing is caused by velocity variations due to changes in pore size and shape, path tortuosity, dead ends, obstructions, and path connectivity. At the macroscopic level, mixing is caused by velocity variations that accounted for complexities of path lines due to local variability of hydraulic conductivity. At the megascopic scale, mixing is occurred by variations in velocity field due to distinct variations in permeability contrast from one lithology to another.

Three types of mixing mechanisms are considered in the literature [Sahimi, 1993]. The transport behaviour can be classified into Fickian transport, diffusive transport in which the variance grows linearly with travel time (an example of this type is dispersion in homogeneous fields), or non-Fickian (anomalous) transport in which the concentration gradient is not directly proportion to the solute mass flux. The non-Fickian behaviour is linked to heterogeneous media by several researchers [Schwartz, 1977; Matheron and de Marsily, 1980; Neuman, et al., 1987; Dagan, 1984, 1994]. This non-Fickian regime exists when the dispersion process is time or space dependent. The non-Fickian can also be classified to be a sub-diffusive transport (fractal dispersion) in which the growth of the spreading variance is slower than linearly with time, or a super-diffusive transport (hyperdiffusion) in which the evolution of the spreading variance growth faster than linear with time (an example of this type is hydrodynamic convection, such as transport in a layered medium with no pore scale transverse dispersivity). These mechanisms can be manifested by considering longitudinal variance in time:

$$\sigma^2_{xx} \sim t^\alpha \tag{6.17}$$

$$D_{xx} \sim t^{\alpha-1} \tag{6.18}$$

where, σ^2_{xx} is the variance in longitudinal direction, D_{xx} is the longitudinal dispersion coefficient, t is elapsed time from starting injection and α is an exponent describing the transport regime. For diffusive dispersion, $\alpha=1$, $\alpha<1$ for sub-diffusive dispersion (occurs in systems with geometric constraints such as fractals), and $\alpha>1$ for super-diffusive dispersion (enhanced diffusion which is typical for chaotic dynamics and turbulence e.g. $\alpha=2$ in case of hydrodynamic convection such as transport in layered system with no local scale transverse dispersivity). Fig.(6.2) shows a sketch of these different types. The presented experiments show these various types of mixing mechanisms and provide the controlling conditions under which these types of mixing occur.

6.8 Outcomes of The Experiments

The results of the experiments can be expressed in terms of statistical moments. These moments are estimated from the particle-cloud [Tompson, et al., 1990] and their mathematical expressions are presented below:

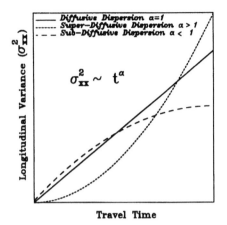 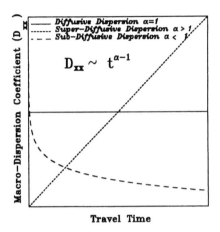

Fig.(6.2) Graphical Illustration of The Various Mixing Mechanisms.

(i) Evolution of The Centroid Displacement
The centroid location is given by,

$$X_c(t) = \frac{1}{N_p} \sum_{k=1}^{N_p} X_k(t) \qquad (6.19)$$

$$Y_c(t) = \frac{1}{N_p} \sum_{k=1}^{N_p} Y_k(t) \qquad (6.20)$$

where, $X_k(t), Y_k(t)$ are the x and y coordinates of particle k at time t.

(ii) Evolution of The Displacement Variances
The spread around the centroid is given by the second moments as,

$$\sigma_{xx}^2(t) = \frac{1}{N_p} \sum_{k=1}^{N_p} \left[X_k(t) - X_c(t) \right]^2 \qquad (6.21)$$

in x-direction, and similarly the spread in y-direction is,

165

$$\sigma_{yy}^2(t) = \frac{1}{N_p} \sum_{k=1}^{N_p} \left[Y_k(t) - Y_c(t) \right]^2 \tag{6.22}$$

(iii) Evolution of The Plume Angle of Rotation
First, the cross-covariance is estimated by,

$$\sigma_{xy}^2(t) = \frac{1}{N_p} \sum_{k=1}^{N_p} \left[X_k(t) - X_c(t) \right]\left[Y_k(t) - Y_c(t) \right] \tag{6.23}$$

and consequently the angle of rotation, θ, measured with respect to the x-axis is given by

$$\theta = \frac{1}{2} \tan^{-1} \left[\frac{2\sigma_{xy}^2}{\sigma_{xx}^2 - \sigma_{yy}^2} \right] \tag{6.24}$$

Higher-order moments can also be calculated to estimate skewness and kurtosis of a particle cloud. These moments will be calculated when necessary during the experiments to test if concentration distributions are Gaussian or not.

(iv) Macro-Dispersion Coefficients
Parameters of the transport equation are dependent on the length scale under consideration (scale-dependent parameters). The dispersion coefficient at pore scale, calculated from column experiments, is in order of magnitude of the grain size. At aquifer scale the dispersion coefficient is larger (macrodispersion). Macroscopic dispersion is caused by hydraulic conductivity contrasts within a medium. The macrodispersion coefficient is defined as half of the rate of change of the second spatial moments of the plume. A necessary condition to define a constant macrodispersion coefficient and consequently a unique macrodispersivity of the medium is that the spatial variance of the particles increases linearly with time. Then the rate of change of the spatial variance is constant and can be related to the macodispersion coefficient as [Dagan, 1982],

$$D_{ij} = \frac{1}{2} \frac{d\sigma_{ij}^2}{dt} \qquad\qquad i, j = x, y \tag{6.25}$$

This coefficient is a function of time t (or of the travel distance from the input zone) if σ^2_{xx} does not grow linearly with time leading to non-Fickian regimes. For estimation of dispersion coefficient during the experiments the following expression is used,

$$D_{ij} \approx \frac{\sigma^2_{ij}(t+\Delta t) - \sigma^2_{ij}(t)}{2 \Delta t} \qquad i, j = x, y \qquad (6.26)$$

(v) Spatial Concentration Distributions
Solving the advection dispersion equation, Eq.(6.1), by Eulerian methods leads to a computed concentration at discrete points in the grid. The random walk method does not produce concentration values. The method produces discrete particle displacements indicating the presence of a certain mass in a grid cell. The tracer concentration within the flow domain is determined by superimposing a rectangular grid of dimensions Δx, and Δy in x and y direction respectively and calculating the probability distribution of the particles. The grid averaged concentration is given by,

$$C_{ijk} = \frac{M_o}{\varepsilon \Delta x \Delta y} f_{ijk} \qquad (6.27)$$

where, C_{ijk} is the volume-averaged concentration in grid ij at time $k\Delta t$, M_o is the released mass, and f_{ijk} is the probability that a particle located at the source at time zero will be found in gridblock ij at time $k\Delta t$.
The quantity f_{ijk} is calculated in the model using the relation,

$$f_{ijk} = \frac{n_{ijk}}{N_p} \qquad (6.28)$$

where, n_{ijk} is the number of particles in element ij at time $k\Delta t$, and N_p is the total number of particles used to discretize the source.

(vi) Breakthrough Curves
In some situations it is convenient or required to monitor the temporal changes of the concentration at fixed locations, the so-called breakthrough curves, rather than determining the complete spatial distribution of the concentration field. A distinction between the volume-averaged concentration, C, defined in Eq.(6.27) and flux-averaged concentration, c_f, is important. The flux-averaged concentration, c_f, is defined as the ratio of the specific solute mass flux, S_x, to the specific fluid flux, q_x, [Kreft and Zuber, 1978],

$$c_f = \frac{S_x}{q_x} \tag{6.29}$$

In case of one-dimensional solute transport equation, the flux-averaged concentration and the volume-averaged concentration are related by [Parker and van Genuchten, 1984],

$$c_f = C - \frac{D_{xx}}{V_x} \frac{\partial C}{\partial x} \tag{6.30}$$

The breakthrough curves are evaluated in the model by counting the particles that cross a prespecified section perpendicular to the mean flow direction (x-direction). The breakthrough curves are transformed to cumulative breakthrough curves by successive summation of the particles that crossed the section.

Three sections are chosen at different distances from the source. The cumulative breakthrough curves can show the Gaussianity or non-Gaussianity of the concentration distribution. Another test of Gaussianity requires the computation of higher-order statistical moments of the concentration field which in turn depends on triple moment (to estimate skewness) and quadruple moment (to estimate kurtosis), etc. In some cases breakthrough curves, which are not computational demanding, or higher-order moments are evaluated.

6.9 Testing The Transport Model

6.9.1 Test Case 1: Homogeneous Formation with Point Source

The two-dimensional random walk transport model 'RWM2D' is tested for some academic cases for which analytical solutions exist. The first case is the propagation of an instantaneous tracer pulse in a steady state uniform flow in homogenous medium with $K = 10$ m/day, $\alpha_l=0.5$m and $\alpha_t= 0.05$m. The analytical solution of the concentration field is [Bear, 1972],

$$C(x,y,t) = \frac{M_o/(\varepsilon \, H)}{\sqrt{4\pi \alpha_l V_x t} \sqrt{4\pi \alpha_t V_x t}} \exp\left[-\left[\frac{(x-X_o-V_x \, t)^2}{4\alpha_l V_x t} + \frac{(y-Y_o)^2}{4\alpha_t V_x t} \right] \right] \tag{6.31}$$

where, H is the aquifer thickness.

The growth of the spatial moments with time is given by,

$$
\begin{aligned}
X_c(t) &= X_o + V_x t \\
Y_c(t) &= Y_o + V_y t \\
\sigma^2_{xx}(t) &= 2D_{xx} t \\
\sigma^2_{yy}(t) &= 2D_{yy} t
\end{aligned}
\tag{6.32}
$$

where, (X_o, Y_o) are the source coordinates at time $t=0$.
The dispersion coefficients are defined as,

$$
\begin{aligned}
D_{xx} &= \alpha_l V_x \\
D_{yy} &= \alpha_t V_x
\end{aligned}
\tag{6.33}
$$

The normalized cumulative breakthrough curves at certain control planes can be calculated analytically by integrating the mass flux over y-direction which leads to the one-dimension solution [Bear, 1972],

$$
\frac{N_{cp}(t)}{N_p} = 1 - \frac{1}{2} \, erfc \left[\frac{x_{cp} - X_o - V_x t}{2\sqrt{D_{xx} t}} \right]
\tag{6.34}
$$

where, x_{cp} is the x-coordinate of the control plane, $N_{cp}(t)$ is the cumulative counting of particles crossed the control plane at time t, and $erfc$ is the complementary error function.

Fig.(6.3) and Fig.(6.4) show the results of test case 1. The results are fairly good in terms of plume spatial moments (Fig.(6.3)) and concentration fields (Fig.(6.4)). The Fickian regime is quite clear from the results. The breakthrough curves show Gaussianity as expected from Eq.(6.34).

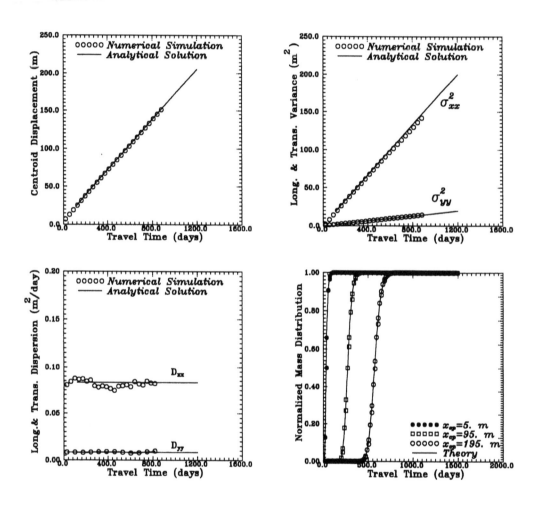

Fig.(6.3) Plume Spatial Moments and Breakthrough Curves in Test Case 1 (Homogeneous Field).

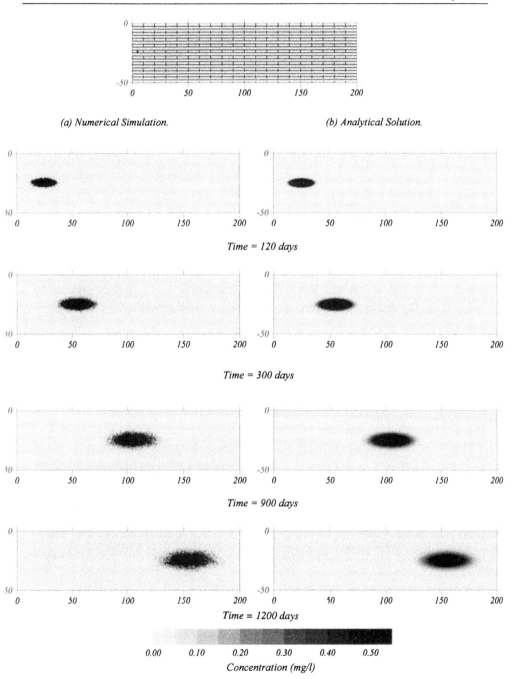

(a) Numerical Simulation. *(b) Analytical Solution.*

Time = 120 days

Time = 300 days

Time = 900 days

Time = 1200 days

0.00   0.10   0.20   0.30   0.40   0.50

Concentration (mg/l)

Fig.(6.4) Plume Evolution in Test Case 1 (Homogeneous Field) (a) Numerical and (b) Analytical Results.

171

6.9.2 Test Case 2: Perfectly Layered Formation

The second test is a propagation of a tracer introduced by a line source in a perfectly layered aquifer composed of two geological materials with K_1=5m/day, K_2=10m/day, $\alpha_l=\alpha_t=0$ (pure convection) and with flow parallel to the bedding. Fig.(6.5) shows the pattern used in this test. The first and second spatial moments of the particle cloud and macro-dispersion coefficient are [Mercado, 1967],

$$X_c(t) = X_o + \langle V_x \rangle\, t$$

$$\sigma^2{}_{xx}(t) = \left[\frac{\sigma_K}{\langle K \rangle}\right]^2 \langle V_x \rangle^2\, t^2 \qquad (6.35)$$

$$D_{xx}(t) = \left[\frac{\sigma_K}{\langle K \rangle}\right]^2 \langle V_x \rangle^2\, t$$

where, $\langle K \rangle$ is the arithmetic mean of the conductivity, $\sigma^2{}_K$ is the variance in K, and $\langle V_x \rangle$ is the arithmetic mean of the velocity field given by,

$$\langle V_x \rangle = \frac{\langle K \rangle}{\varepsilon} J_x \qquad (6.36)$$

where, J_x is the gradient in the mean flow direction (x-direction)

Fig.(6.5) shows a comparison of test case 2 with the numerical simulation. The results also show good agreements for the spatial moments. Concentrations are not displayed because each cell in the field is either contaminated with $C=C_o$ or not contaminated $C=0$.

Eq.(6.35) and its graphical representation Fig.(6.5) demonstrate that the longitudinal dispersion coefficient grows without bounds indicating a super-diffusive regime with power $\alpha=2$. This growth of the dispersion coefficient is the 'scale-dependent dispersion' and is due to the heterogeneity of a pure advective transport process.

6.10 Interpretation and Discussion of The Experimental Results

6.10.1 Influence of Single Heterogeneity (Inclusion)

In these experiments, an inclusion is considered to get an insight in the afore-mentioned mixing mechanisms (section 6.9). Two types of inclusions are investigated. High permeable inclusion (with permeability of the inclusion K_i=10 m/day

172

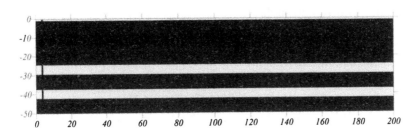

Perfectly Layered Formation.

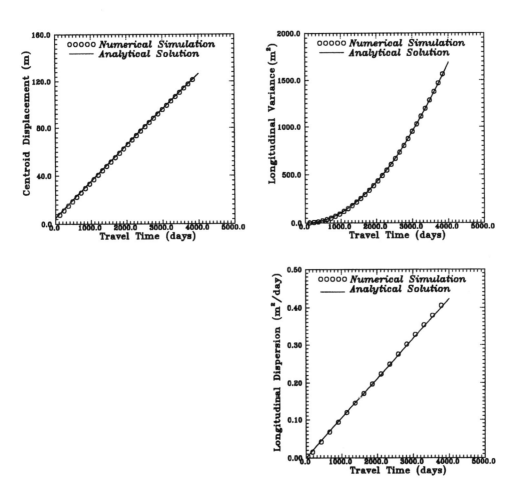

Fig.(6.5) Perfectly Layered Medium in Test Case 2: Analytical and Numerical Plume Spatial Moments.

173

and permeability of the surrounding medium K_m=1m/day) and a low permeable one (K_i=1 m/day and K_m=10 m/day).

Fig.(6.6a) and Fig.(6.6b) show the flow field, in terms of a flownet, and the transport field, in terms of plume evolutions. The longitudinal and transverse spatial moments and dispersion coefficients are presented in Fig.(6.7) for both cases. Inspection of these figures show that due to the heterogeneity flow may advance (in case of converging flow lines) or retard (in case of diverging flow lines) the flow along the streamlines. As a consequence the transport behaviour may change from normal to anomalous diffusion or vice versa.

One may recognize in Fig.(6.7) an actual decrease in the transverse variance leading to negative dispersion. This is due to the presence of converging flow lines towards the inclusion, which squeeze the plume laterally and stretch it in the longitudinal direction. All kinds of mixing regimes (diffusive, subdiffusive and superdiffusive) can be seen from this example. In real aquifers, however, more than one inclusion exists and their influences interfere in a more complex fashion. The following experiments will show the behaviour of more complicated cases.

6.10.2 Markovian Field with Low and High Contrast in Conductivity

In these experiments, the influence of a contrast in conductivity is investigated. A two-dimensional Markovian field has been generated with four different geological materials. The generated field is characterized by the transition probabilities given in Table (6.1). Two cases are considered: case 1, corresponds to low contrast in conductivity and case 2, corresponds to high contrast in conductivity. The corresponding conductivities in both cases are displayed in Table (6.2). The results of both cases in terms of plume evolutions are displayed in Fig.(6.8) and Fig.(6.9) respectively.

Visual detection of the simulations (Fig.(6.8) and Fig.(6.9)) show that the plume is moving slower in case of high contrast than in case of low contrast and the spreading is covering more area in case of high contrast than in case of low contrast. This is due to the existence of low permeable zones where the particles may move slowly towards these zones or caught up there causing elongated tails.

The plume spatial moments are plotted in Fig.(6.10) for both cases. The retardation of the plume mentioned above manifests clearly in the first spatial moment (centroid displacement). The longitudinal variance, Fig. (6.10c) shows larger spreading in case of high contrast than in case of low contrast. The macrodispersion, Fig.(6.10d) shows a higher value of the macrodispersion coefficient in case of high contrast. It seems also that the plume has reached asymptotic macrodispesion within 400 days in case of low contrast. However, this asymptotic macrodispersion is not yet obtained in case of high contrast.

The breakthrough curves, Fig.(6.10b), show elongated tails (non-Gaussianity) in case of high contrast. It seems that the breakthrough curves tend to become

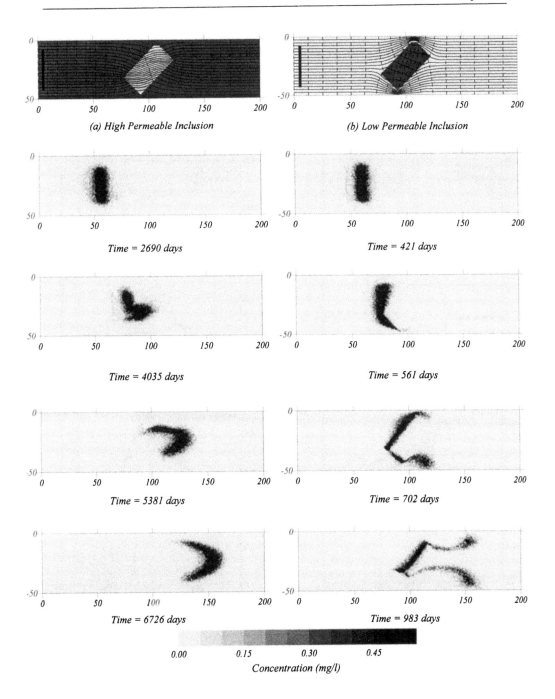

(a) High Permeable Inclusion

(b) Low Permeable Inclusion

Time = 2690 days

Time = 421 days

Time = 4035 days

Time = 561 days

Time = 5381 days

Time = 702 days

Time = 6726 days

Time = 983 days

0.00   0.15   0.30   0.45

Concentration (mg/l)

Fig.(6.6) Plume Evolution in Case of Single Inclusion: (a) High Permeable Inclusion. (b) Low Permeable Inclusion.

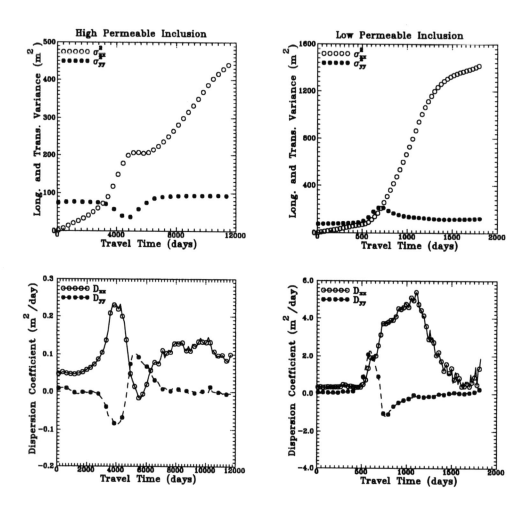

Fig.(6.7) Spatial Moments of The Plume in Case of Single Inclusion.

Table (6.1) Statistics of Fig.(6.8) and Fig.(6.9).

Length of The Given Section (m)=   300.
Depth of The Given Section (m)=   50.
Sampling interval in X-axis (m)=   3.
Sampling interval in Y-axis (m)=   2.

Input Statistics   Calculated Statistics

Horizontal Transition Probability Matrix

State	1	2	3	4	State	1	2	3	4
1	0.980	0.050	0.050	0.100	1	0.851	0.044	0.037	0.067
2	0.100	0.980	0.050	0.050	2	0.114	0.808	0.051	0.027
3	0.100	0.050	0.980	0.050	3	0.103	0.030	0.817	0.050
4	0.100	0.050	0.050	0.980	4	0.099	0.044	0.036	0.821

Vertical Transition Probability Matrix

State	1	2	3	4	State	1	2	3	4
1	0.400	0.200	0.200	0.200	1	0.540	0.121	0.145	0.194
2	0.200	0.400	0.200	0.200	2	0.330	0.358	0.126	0.186
3	0.200	0.200	0.400	0.200	3	0.189	0.145	0.388	0.278
4	0.200	0.200	0.200	0.400	4	0.331	0.170	0.189	0.310

Table(6.2) Conductivities of The States used in Markovian Fields.

State	Colour on The Map	Low Contrast	High Contrast
1	Very Light Grey	80. m/day	100. m/day
2	Light Grey	50. m/day	10.0 m/day
3	Dark Grey	20. m/day	1.00 m/day
4	Black	10. m/day	0.10 m/day

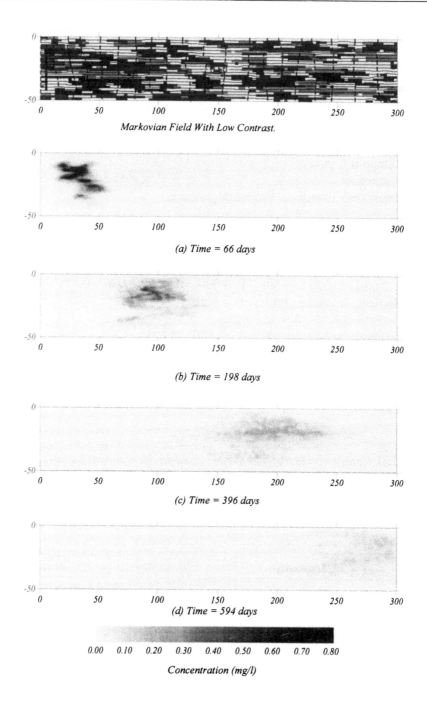

Markovian Field With Low Contrast.

(a) Time = 66 days

(b) Time = 198 days

(c) Time = 396 days

(d) Time = 594 days

0.00 0.10 0.20 0.30 0.40 0.50 0.60 0.70 0.80

Concentration (mg/l)

Fig.(6.8) Plume Evolution in Markovian Field With Low Contrast in Conductivity.

178

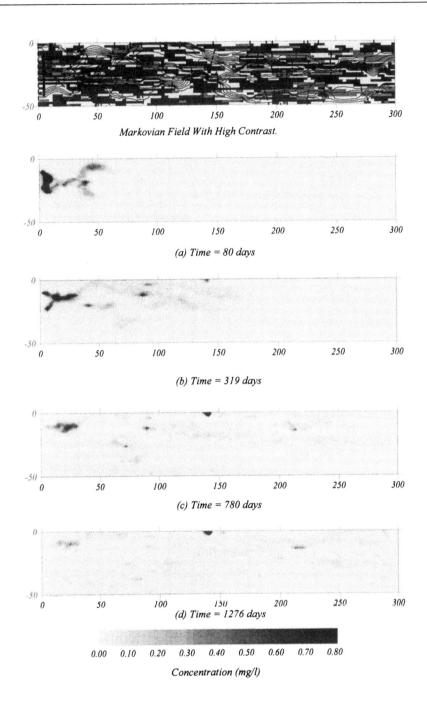

Markovian Field With High Contrast.

(a) Time = 80 days

(b) Time = 319 days

(c) Time = 780 days

(d) Time = 1276 days

0.00   0.10   0.20   0.30   0.40   0.50   0.60   0.70   0.80

Concentration (mg/l)

Fig.(6.9) Plume Evolution in Markovian Field With High Contrast in Conductivity.

179

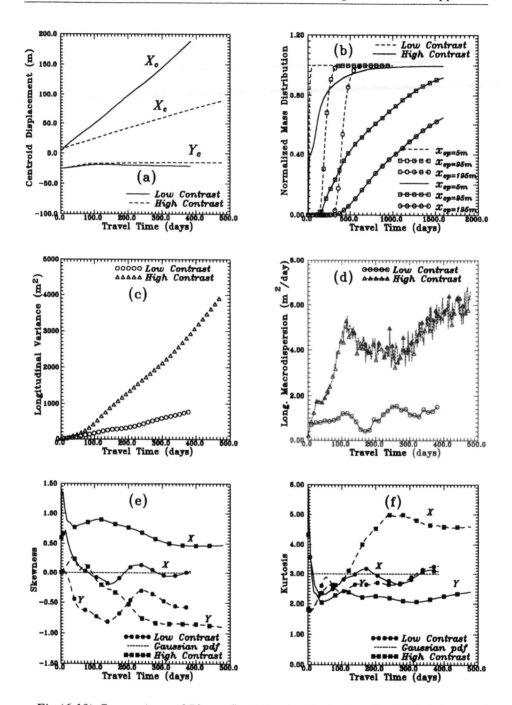

Fig.(6.10) Comparison of Plume Statistics in Markovian Field With Low and High Contrast in Conductivity.

Gaussian in case of low contrast. This behaviour is confirmed when the higher order moments, skewness and kurtosis in both *x* and *y* directions, are considered. Therefore, the moments are displayed in Fig.(6.10e) and Fig.(6.10f) respectively. The figures show that the concentration distributions are approaching Gaussian distributions in case of low contrast within about 400 days except the skewness in *y*-direction which is due to the top boundary. The plume is directed over there because of preferred flow paths whereas, in case of high contrast the plume is still far from Gaussianity within the time period of the experiments.

6.10.3 Effect of Various Characterization Methodologies (Markovian versus Gaussian)

The influence of various types of characterization methodologies on flow and transport behaviour are considered. The commonly used stationary Gaussian fields and the Markovian fields are investigated. The statistical equivalence between the two methods is made by the following procedure. A geometrical structure has been generated by the Markov model with prescribed transition probability matrices and sampling intervals given in Table (6.1). The outcome of the simulation is displayed in Fig.(6.8). To each unit in the geological system a hydraulic conductivity has been assigned according to Table (6.2). The spatial mean, variance and correlation lengths of the generated realization are computed by fitting an anisotropic exponential autocorrelation function to the calculated autocorrelation from the Markovian field. Fig. (6.11) shows both experimental and fitted autocorrelation used in the calculations. The estimated parameters are displayed on Table (6.3).

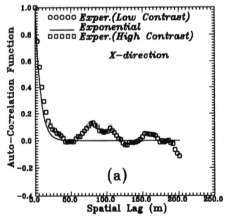 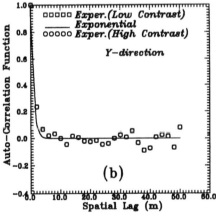

Fig.(6.11) Fitting Exponential Auto-Correlation Function to Markovian Fields with Various Contrasts.

Table(6.3) Conductivity Statistics from The Generated Markovian Field.

Parameter	Low Contrast	High Contrast
K_a	47.1 m/day	41.9 m/day
σ_K	29.4 m/day	47.6 m/day
cv	0.62	1.14
λ_x	6.97 m	6.84 m
λ_y	0.88 m	0.82 m

These parameters are used to generate stationary Gaussian random field with the turning bands method. Then, a flow and transport have been calculated in both fields with given identical boundary and initial conditions.

Fig.(6.12) shows the marginal probabilities and *PDF* for both Markovian and Gaussian fields under low and high contrast conditions respectively. The results of this test in terms of flownets and plume evolutions are shown in Fig.(6.13) and Fig.(6.14) for low and high contrasts. This test may be comparable with its statistically equivalent Markovian field for low and high contrast displayed in Fig.(6.8) and Fig.(6.9).

The plume spatial moments in both low and high contrast for Gaussian and Markovian fields are shown in Fig.(6.15) and Fig.(6.16). In case of low contrast, Fig.(6.15), the results show that plume centroid, Fig.(6.15a), is moving slower in Markovian field than in Gaussian field but the differences are not significant in contrary to the case of high contrast, Fig.(6.16a). The breakthrough curves, Fig.(6.15b), show Gaussianity in both Gaussian and Markovian fields with low contrast. Whereas, in case of high contrast the non-Gaussianity (elongated tails) is clear from the results. The reason for that is the existence of low permeable zones (0.1 m/day) near the source where the particles are moving very slowly until they are released.

Fig.(6.15c) shows the longitudinal spreading of the plume which is larger in Markovian field than in Gaussian field especially in case of high contrast Fig.(6.16c).

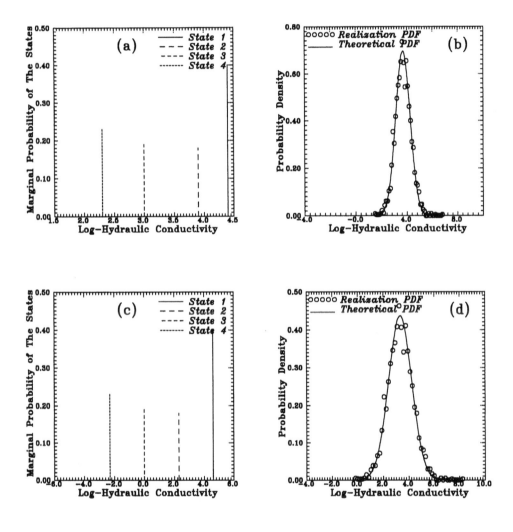

Fig.(6.12) PDFs of Log-Conductivity in Case of Markovian Fields (a,c) and Their Equivalent Gaussian Fields (b,d).

183

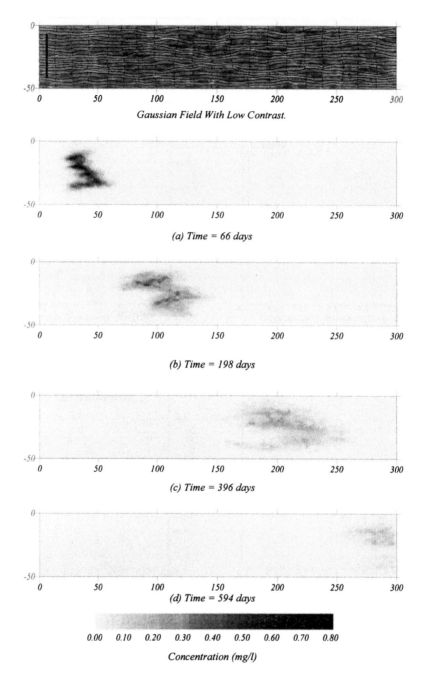

Gaussian Field With Low Contrast.

(a) Time = 66 days

(b) Time = 198 days

(c) Time = 396 days

(d) Time = 594 days

Concentration (mg/l)

Fig.(6.13) Plume Evolution in Gaussian Field Equivalent to Markovian Field in Fig.(6.8) With Low Contrast in Conductivity.

184

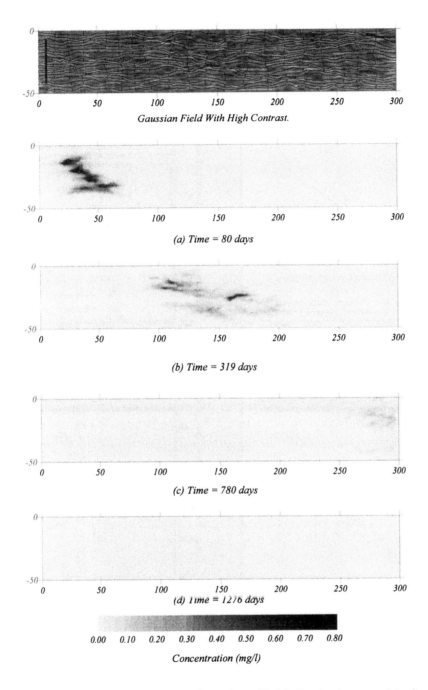

Gaussian Field With High Contrast.

(a) Time = 80 days

(b) Time = 319 days

(c) Time = 780 days

(d) Time = 1276 days

Concentration (mg/l)

Fig.(6.14) Plume Evolution in Gaussian Field Equivalent to Markovian Field in Fig.(6.9) With High Contrast in Conductivity.

185

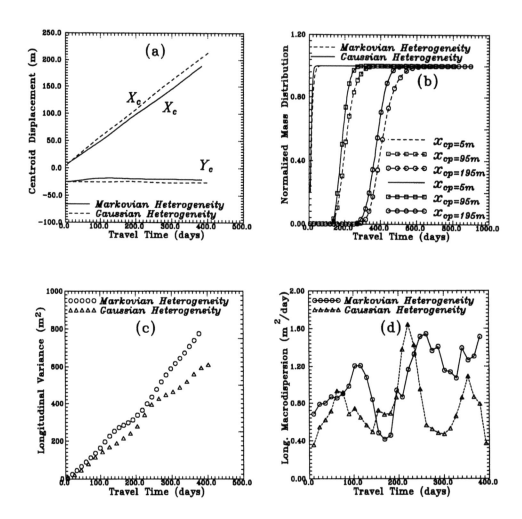

Fig.(6.15) Comparison of Plume Statistics in Case of Markovian Field With Its Equivalent Gaussian Field With Low Contrast in Conductivity.

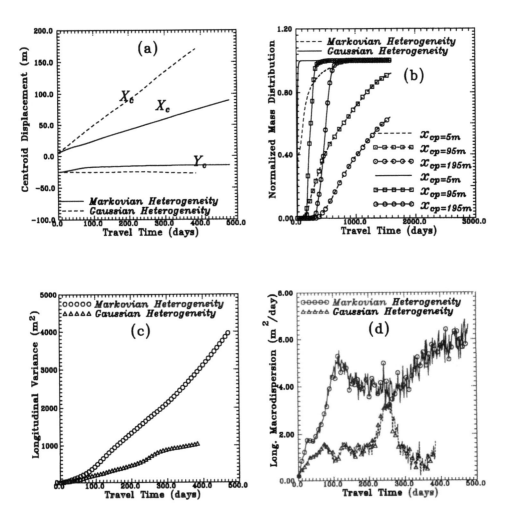

Fig.(6.16) Comparison of Plume Statistics in Case of Markovian Field With Its Equivalent Gaussian Field With High Contrast in Conductivity.

187

The reason is the discrete nature of Markovian heterogeneity. The plume faces sharp contrasts in conductivity between adjacent cells that causes high degree of spreading. However, in Gaussian field the contrast between the adjacent cells are smooth due to the influence of continuous autocorrelation.

Fig.(6.15d) shows the macrodispersion coefficient in case of low contrast. The curves show an oscillation in the magnitude of the dispersion with some upward trend indicating an increase of the dispersion coefficient due to heterogeneity. Both curves, globally, indicate more or less the same behaviour, whereas, in case of high contrast, Fig.(6.16d) shows that there is a quite distinction both in the magnitude and the global trend. The Gaussian filed seems to reach an asymptotic average value of about 2.0 m^2/day while the Markovian field does not seem to reach an asymptotic value within the time of the experiment but if an average value could be estimated it would be around 6.0 m^2/day which is higher than the corresponding Gaussian field.

6.10.4 Effect of Two Scales of Variability (Compound Fields)

The effects of two scales of variability on transport characteristics are investigated. In this sort of experiments the aquifer heterogeneity, as represented by the hybrid model is used for this test. A megascopic geological structure is synthesized by Markov model with four different geological units and with transition probabilities given in Table (6.4). Auto-correlated conductivity values drawn from a log-normal distributions are assigned to each geological units by a turning bands algorithm to represent macroscopic variability. Table (6.5) shows the parameters used for set 1 of the experiments.

Flow and transport are solved numerically under two conditions the first is geological structure without internal variability (pure Markovian field) and the second is geological structure with internal variability (compound field).

Another set of this sort of experiments is considered to get more insight about the transport behaviour in these fields. An inversion of the fields, mentioned in set 1, is considered to represent set 2. The corresponding statistics are displayed in Table (6.6). *PDFs* of both sets are displayed in Fig.(6.17). Flow and transport results in terms of flownets superimposed over the structures and plume evolutions are displayed in Fig.(6.18) and Fig.(6.19) for set 1 and in Fig.(6.21) and Fig.(6.22) for set 2. The spatial moments in set 1 and set 2 are displayed in Fig.(6.20) and Fig.(6.23) respectively.

Table (6.4) Statistics of Fig.(6.18).

Length of The Given Section (m)=   300.
Depth of The Given Section (m)=   50.
Sampling interval in X-axis (m)= 2.
Sampling interval in Y-axis (m)= 1.

Input Statistics   Calculated Statistics

Horizontal Transition Probability Matrix

State	1	2	3	4	State	1	2	3	4
1	0.960	0.010	0.010	0.020	1	0.975	0.007	0.002	0.016
2	0.010	0.970	0.010	0.010	2	0.008	0.977	0.005	0.010
3	0.010	0.020	0.960	0.010	3	0.024	0.023	0.948	0.005
4	0.010	0.010	0.010	0.970	4	0.022	0.007	0.007	0.965

Vertical Transition Probability Matrix

State	1	2	3	4	State	1	2	3	4
1	0.930	0.010	0.030	0.030	1	0.922	0.012	0.018	0.048
2	0.100	0.800	0.050	0.050	2	0.072	0.880	0.027	0.021
3	0.050	0.050	0.800	0.100	3	0.024	0.125	0.732	0.119
4	0.050	0.050	0.030	0.870	4	0.040	0.041	0.006	0.913

Table (6.5) Conductivity Statistics in Set 1.

Parameter State	w_i	$\langle K \rangle$ m/day	σ_K m/day	$\langle Y \rangle$	σ_Y	λ_x (m)	λ_y (m)
1	0.34	100.	100.	4.3	0.83	20.	5.
2	0.26	10.	10.	1.96	0.83	20.	5.
3	0.10	1.0	1.0	-0.35	0.83	20.	5.
4	0.30	0.1	0.1	-2.65	0.83	20.	5.

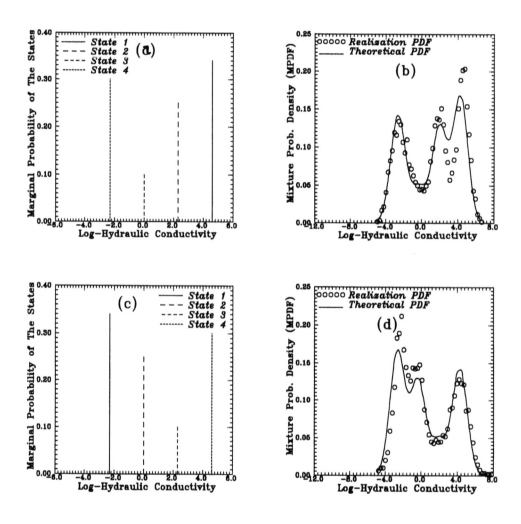

Fig.(6.17) Marginal Probabilities and Mixture PDFs of Log-Conductivity in Set 1 (a,b) and Set 2 (c,d).

Table (6.6) Conductivity Statistics in Set 2 (Inversion of Set 1).

Parameter State	w_i	$\langle K \rangle$ m/day	σ_K m/day	$\langle Y \rangle$	σ_Y	λ_x (m)	λ_y (m)
1	0.34	0.1	0.1	-2.65	0.83	20.	5.
2	0.26	1.	1.	-0.35	0.83	20.	5.
3	0.10	10.	10.	1.96	0.83	20.	5.
4	0.30	100.	100.	4.3	0.83	20.	5.

In order to interpret the results of these experiments, one may decompose transport into two parts, a convective part which takes place through the preferential flow paths (highly correlated channels) and a dispersive part which takes place due to local scale heterogeneity. The superdiffusion regime is due to convection. In this regime the tracer finds permeable zones (channels, pathways, etc.) as it travels through the systems. This behaviour is clear in Fig.(6.18). Most of the particles enter the high conductivity zone near the source (preferential flow path) leading to a convection dominant transport and a superdiffusive regime. This behaviour depends upon the existence of these preferred paths in the generated field and whether or not the tracer particles can migrate into that zones. Other particles may move more slowly towards low permeable zones or caught up in these zones as shown in Fig.(6.18). A considerable time is needed for the particles to cross these low permeable zone and to get released into the high permeable zones.

The results of these tests, displayed in Fig.(6.20a,c) and Fig.(6.23a,c), show that the plumes are slower in case of introducing internal variability (compound field) than in case of without internal variability (pure Markovian field). This behaviour reflects the influence of log-normal distribution used to generate the internal variability. The values of conductivity higher than the arithmetic mean is less likely to exist in the field than the values of conductivity less than the arithmetic mean. This means that the aquifer units are dominated by conductivities less than the arithmetic mean. Due to that fact, the transport is slower.

Plume statistics of set 1 displayed in Fig.(6.20d), show that in absence of internal variability, the macrodispersion coefficient grows without bounds indicating superdiffusive regime within the time limit of the experiment. If the internal variability is present, the macrodispersion coefficient reaches an asymptotic value

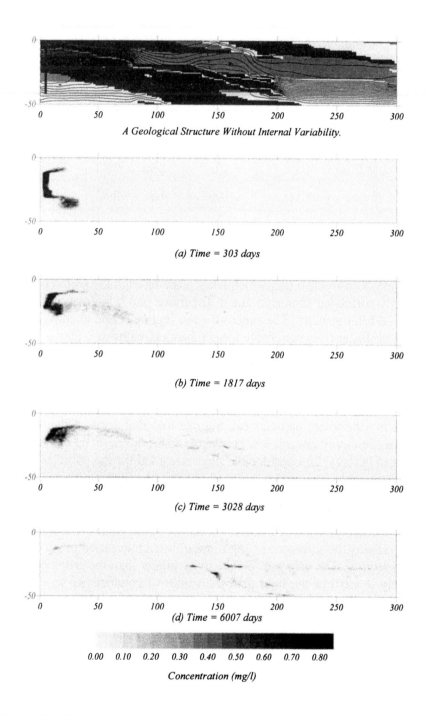

A Geological Structure Without Internal Variability.

(a) Time = 303 days

(b) Time = 1817 days

(c) Time = 3028 days

(d) Time = 6007 days

0.00   0.10   0.20   0.30   0.40   0.50   0.60   0.70   0.80

Concentration (mg/l)

Fig.(6.18) Plume Evolution in Structure Without Internal Variability: Set 1.

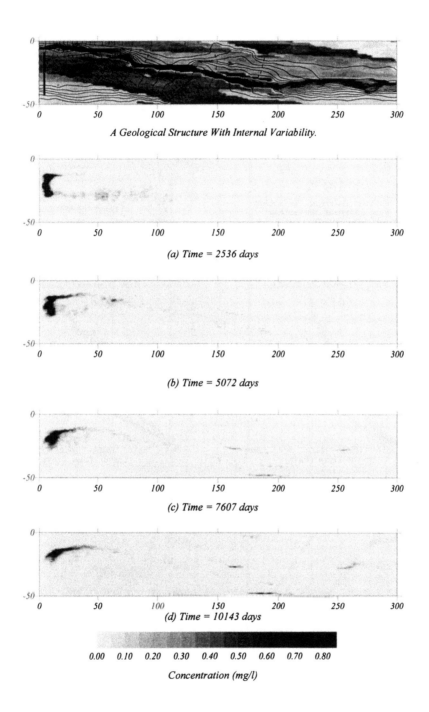

A Geological Structure With Internal Variability.

(a) Time = 2536 days

(b) Time = 5072 days

(c) Time = 7607 days

(d) Time = 10143 days

Concentration (mg/l)

Fig.(6.19) Plume Evolution in Structure With Internal Variability: Set 1.

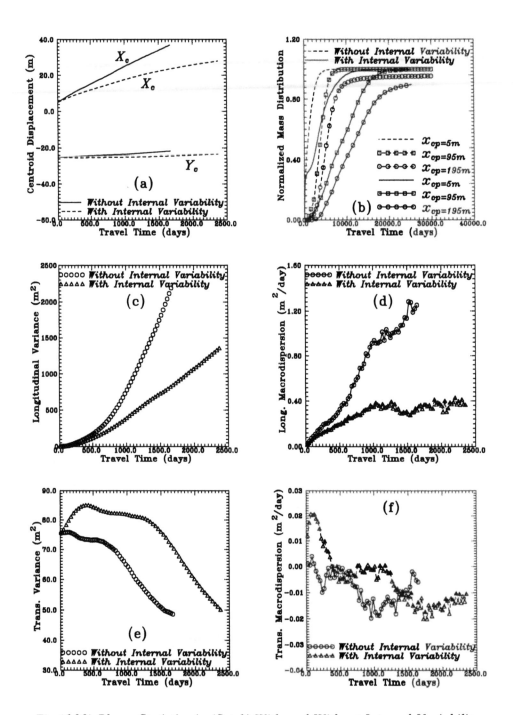

Fig.(6.20) Plume Statistics in 'Set 1' With and Without Internal Variability.

within approximately the same time period. This indicates that a diffusive regime has been reached. This behaviour can be accounted for by the internal variability which enhances the mixing process and leads to an asymptotic dispersion coefficient.

Fig.(6.20b) shows comparisons of the breakthrough curves in both cases (with and without internal variability in set 1). These curves can give some impressions about the long time behaviour till 30,000 days. The concentration distributions for both cases are still far from Gaussian.

On the other hand, plume statistics of set 2 presented in Fig.(6.23d) shows that in case of absence of the internal variability, the macrodispersion coefficient grows until it reaches more or less an asymptotic value within the time limit of the experiment. Whereas, with presence of the internal variability the macrodispersion fluctuates introducing different transport regimes within the same time period. One may distinguish different stages. In the first stage, an asymptotic value of macrodispersion coefficient has been reached after about 200 days from the time of injection indicating a diffusive regime. A second stage appears, where the plume goes faster in the high permeable zone (superdiffusive regime) following the preferential flow paths within that zone till it reaches another asymptotic dispersive regime after 800 days. Later on, the macrodispersion coefficient decreases due to divergence of the flow lines at the low permeable zone indicating a subdiffusive regime.

The general behaviour can be explained as follows. In case of the absence of internal variability the only heterogeneity encountered by the plume is that near the source. As long as the plume propagates in the system it gets only in touch with the heterogeneity near the source. Major parts of the plume are advancing with relatively high velocity in the homogeneous highly permeable unit, while other parts of the plume are moving very slowly in the low permeable zones. Since the plume meets only this heterogeneity it gets into a dispersive regime. However, in case of introducing internal variability the plume gets in touch with the heterogeneity in the system in the earlier stage. Later on the plume meets the variability inside the high permeable zone which is moderately high (σ_K=100 m/day). This variability introduces more tortuosity of the flowlines than in the absence of internal variability. As a consequence, it causes the transport regime to change according to the divergence and convergence of the flowlines during the plume propagation in the system. As long as the flow pattern does not repeat itself in a relatively regular manner, the transport process will not be Fickian.

Fig.(6.23b) shows comparisons of the breakthrough curves in both cases (with and without internal variability in set 2). These curves show the long term behaviour until 8000 days. The breakthrough curves in both cases are far from Gaussianity

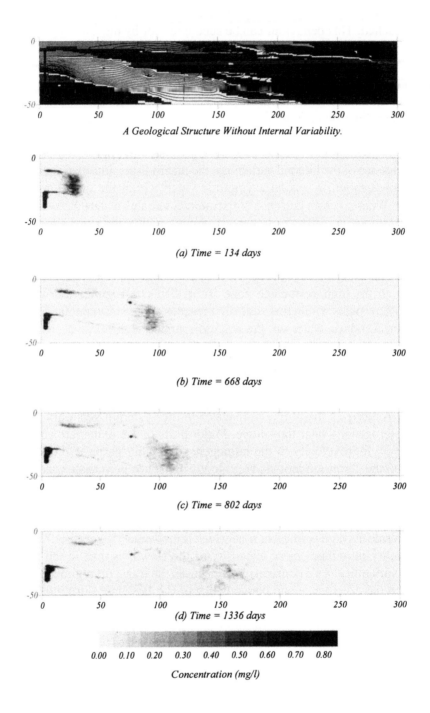

A Geological Structure Without Internal Variability.

(a) Time = 134 days

(b) Time = 668 days

(c) Time = 802 days

(d) Time = 1336 days

Concentration (mg/l)

Fig.(6.21) Plume Evolution in Structure Without Internal Variability: Set 2.

196

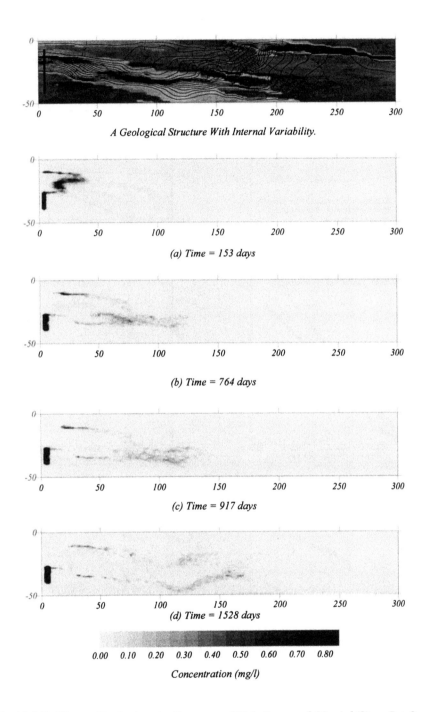

A Geological Structure With Internal Variability.

(a) Time = 153 days

(b) Time = 764 days

(c) Time = 917 days

(d) Time = 1528 days

0.00 0.10 0.20 0.30 0.40 0.50 0.60 0.70 0.80

Concentration (mg/l)

Fig.(6.22) Plume Evolution in Structure With Internal Variability: Set 2.

197

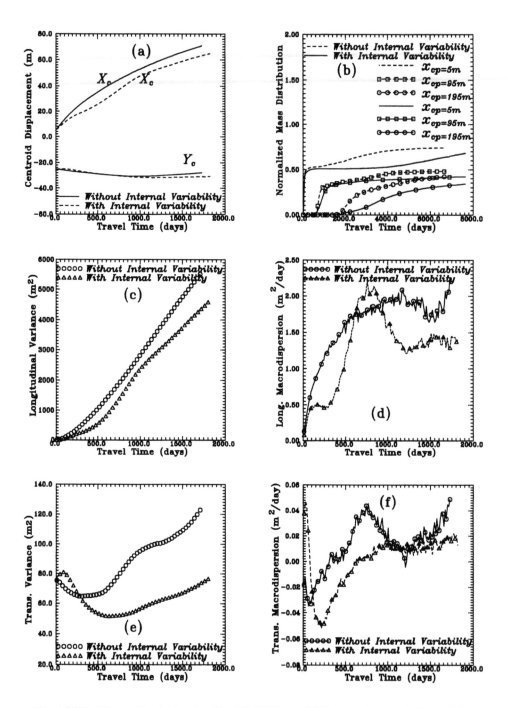

Fig.(6.23) Plume Statistics in 'Set 2' With and Without Internal Variability.

and they show very long tails due to the low permeable zones near the source.

The following conclusion can be stated. The location of the contaminant release zone and the contrast on conductivity near that zone has a significant influence on the long term transport behaviour.

6.10.5 Perfectly Stratified Formation Without Pore Scale Dispersion

Stratifications, as commonly encountered in aquifers, have been given extensive attention in the literature. These stratifications are often governed by regional geological processes which takes place over long time periods. In the literature, these stratifications are treated sometimes as homogeneous layers with constant hydraulic parameters [Marcado, 1967]. However, some fluctuations within the layers (macro-structures) may exist due to transient processes over short time periods during the deposition of these layers.

Therefore, as an experiment, a perfectly stratified formation is considered with and without internal variability. The reason for this test is to investigate the influence of the macroscopic variability in the absence of the pore scale dispersion process and in the case of pronounced superdiffusion. Table (6.7) shows the transition probabilities used to characterize these types of formations. The corresponding conductivity parameters used to generate the internal variability are presented in Table (6.8). Marginal probabilities and *PDFs* corresponding to this case are displayed in Fig.(6.24). Fig.(6.25) displays some results of this test. The flow pattern shows tortuosity in flowlines due to internal variability.

The spatial moments show retardation in the centroid displacement and the growth of longitudinal macodispersion coefficient decreases with time in the presence of internal variability. This is due to the fact that internal variability causes tortuosity of the flowlines between the layers according to the arrangements of high and low permeable zones. This behaviour mimics the influence of a lateral pore scale dispersion process which causes lateral mixing.

Table (6.7) Statistics of Fig.(6.24).

Length of The Given Section (m)=   300.
Depth of The Given Section (m)=   50.
Sampling interval in X-axis (m)=   1.
Sampling interval in Y-axis (m)=   0.5

Input Statistics   Calculated Statistics

Horizontal Transition Probability Matrix

State	1	2	3	4	State	1	2	3	4
1	1.000	0.000	0.000	0.000	1	1.000	0.000	0.000	0.000
2	0.000	1.000	0.000	0.000	2	0.000	1.000	0.000	0.000
3	0.000	0.000	1.000	0.000	3	0.000	0.000	1.000	0.000
4	0.000	0.000	0.000	1.000	4	0.000	0.000	0.000	1.000

Vertical Transition Probability Matrix

State	1	2	3	4	State	1	2	3	4
1	0.600	0.100	0.200	0.100	1	0.538	0.077	0.231	0.154
2	0.100	0.700	0.100	0.100	2	0.049	0.756	0.122	0.073
3	0.150	0.150	0.400	0.300	3	0.111	0.167	0.444	0.278
4	0.100	0.200	0.200	0.500	4	0.071	0.214	0.071	0.643

Table (6.8) Conductivity Statistics in Case of Layered Formation.

Parameter State	w_i	$\langle K \rangle$ m/day	σ_K m/day	$\langle Y \rangle$	σ_Y	λ_x (m)	λ_y (m)
1	0.13	100.	100.	4.26	0.83	10.	2.
2	0.41	10.	100.	-0.05	2.14	10.	2.
3	0.18	1.0	100.	-4.61	3.03	10.	2.
4	0.28	0.1	100.	-9.21	3.72	10.	2.

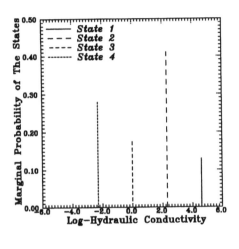

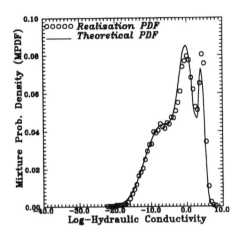

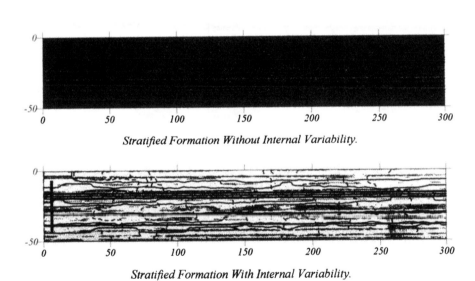

Stratified Formation Without Internal Variability.

Stratified Formation With Internal Variability.

Fig.(6.24) Marginal Probabilities and Mixture PDFs of Log-Conductivity in Perfectly Stratified formation and Flownet.

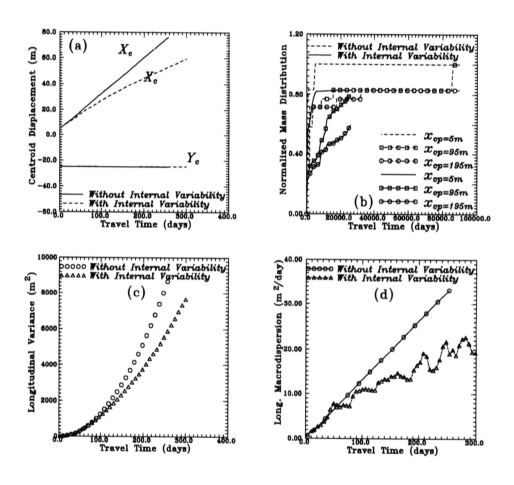

Fig.(6.25) Plume Spatial Moments in Stratified Formation With and Without Internal Variability.

6.11 Summary and Conclusions

The traditional stochastic models for flow and solute transport are based on unimodal stationary log-normally heterogeneous fields. Several researchers have analyzed the flow and transport behaviour under this notion extensively. Only very few studies are devoted to investigate the influence of other methods of heterogeneous characterization. Some geological formations manifest Markovian properties [Krumbein, 1967]. This has motivated the use of a geologically-based type of characterization to describe the spatial structure of the hydraulic properties (e.g. the coupled Markov chain model developed in Chapter 3 and the hybrid model developed in Chapter 5). These methodologies have been used to analyze the significance of uni-modal stationary Gaussian fields, multi-modal non-stationary fields, Markovian and com-pound fields (multiple-scale fields) on the spreading characteristics of contaminated plumes.

In the proposed numerical experiments, single realization hydraulic conductivity fields are generated by these methods and the associated flow and transport problems are solved numerically under specified initial and boundary conditions. Confined aquifer conditions and steady velocity fields are considered. The first two spatial moments of the particle cloud which describe the displacement of the plume centroid and the spread around the centroid, breakthrough curves at some specified locations and macrodispersion coefficients are estimated from the numerical experiments. The various transport regimes (diffusive, superdiffusive and subdiffusive) are observed in these experiments.

From the results, the following conclusive remarks are made:

(1) The numerical experiments give better understanding of the effects of the various characterization methods of conductivity fields on the transport behaviour. For many geological patterns, the concentration fields are no longer Gaussian, particularly when strong heterogeneity and preferential flow paths are present in the system.

(2) The evolution of plume spatial moments and dispersion coefficients in a single realization heterogenous medium do not follow a monotonous increase with time. On the contrary, they go up and down and as a consequence dispersion behaviour displays intermittent chains of Fickian or non-Fickian regimes.

(3) The experiments emphasize that the macrodispersion coefficients are not equivalent to those used at the microscopic level (microdispersion). The higher the contrast in conductivity of the individual units, the larger the value of the macrodispersion coefficient.

(4) The experiments confirm that dispersion, in general, in many geological structures is non-Fickian at least for short to intermediate travel times investigated in these experiments. The existence of high conductivity channels in geological systems

is the main cause of superdiffusive regime, while a fractal (subdiffusive) dispersion regime can be observed in some of the presented formations.

(5) The effects of internal variability within the geological units are more pronounced on transport characteristics than on the hydrodynamics of flow as long as the boundaries between the units are not smeared by the internal variability.

(6) The assumption of log-normal *PDF* of conductivity inside the geological units is reflected in the results, since zones of permeability less than the arithmetic mean of conductivity are more likely to appear in the field than highly permeable zones. This causes retardation in the plume evolution and consequently in its spatial moments.

(7) In stratified formations under absence of pore scale dispersion, the internal variability causes a decrease of the macrodispersion coefficient. This behaviour mimics the influence of pore scale dispersion.

(8) In compound field experiments the internal variability contributes mostly in the mixing processes by introducing tortuosity of the flowlines. This mimics the pore scale dispersion process. In some geological setting the internal variability changes the transport regimes from superdiffusive to diffusive.

(9) It appears that reliable predictions of solute in aquifers requires sufficiently detailed knowledge of the major geological features, specially the preferential flow paths, within the aquifers. It has been shown that the characterization method either Markovian, Gaussian or compound has a strong influence on transport behaviour. The choice of the appropriate method for field applications should be tested carefully. The characterization methodologies proposed in this thesis may help in that respect.

(10) A general conclusion from these experiments is that in order to provide a definitive assessment of concentration fields a deterministic description of the site heterogeneity is required. The single realizations presented here can be viewed as deterministic fields. As a consequence, one may recognize the various mechanisms and complexities in transport processes although convection and dispersion are the only processes concerned in this study. In practical applications, however, a full deterministic description would never be available. Then it should be more realistic to provide assessments of concentration fields in probabilistic context in order to account for uncertainty in site specification. Chapter 7 will focus on that topic.

(11) Laboratory and field studies are needed to assist in model validation. Therefore, some laboratory experiments are carried out described in Chapter 8.

Prediction Uncertainty of Solute Transport in Heterogeneous Formations

7.1 Introduction

Recent field tracer tests of real plumes show that the contaminant plumes are very ragged, see for example, the Twin Lake tracer test [Killey, et al., 1988] and the Bromide plume at Columus Air Force Base [Boggs, et al., 1992]. Plume irregularities and their eratic shape can also be observed from the numerical simulations in the previous chapter. These erratic shapes are accounted for by the heterogeneous structure of natural formations. Therefore, prediction of contaminant spreading in field scale is subjected to high degree of uncertainty due to the uncertainty in the spatial distribution of the hydrological parameters and /or the irregular geological structure of the formations. Due to this fact, it is more realistic to make predictions in a probabilistic context (in terms of mean concentration plus two standard deviations) rather than in the traditional deterministic framework (in terms of mean concentration only).

In stochastic analysis uncertainty in the input parameters is specified in the form of a probability density function, *PDF*, or the first two moments of a distribution. There are several approaches to study the propagation of the uncertainties of the input parameters to the output variables: (1) first-order analysis [Dettinger and Wilson, 1981], (2) spectral-perturbation analysis [Gelhar, 1984], and (3) Monte-Carlo, *MC*, method. The last one is the most widely used tool to determine explicitly and straightforwardly the entire *PDF* of the concentration field and/or its concentration expectation and variance in quite complex flow patterns, boundary conditions and initial disposition of the solute source. The *MC* method has the advantage, of being the most powerful technique in handling parametric uncertainty (which has been extended in this study to non-parametric uncertainty) and it does not suffer from restrictions of linearization, boundary conditions as the analytical solutions presented by Dagan [1984], Gelhar and Axness [1983] and others. It is a feasible approach to problems which are not analytically solvable. However, its drawback is due to intensive computational effort which would become less of concern because of the technical innovations of computer speed. Therefore, the *MC* method is used here. A brief description is presented next.

The main ingredient of *MC* approach is that a large number of equiprobable realizations of the formation structure is generated. Each realization is processed through a transfer function, in our case a deterministic groundwater flow and transport model in order to obtain a series of output responses. The responses of the transfer function are groundwater head, velocity, and concentration fields for each input realization. The envelope curves (95% confidence intervals) of the responses distribution, in our case, the spatial moments of the plume evolution in time, macro-dispersion coefficients, and concentration variance can be estimated by ensemble averaging.

7.2 Set Up of Multi-Realizations Numerical Experiments

A series of two-dimensional numerical experiments is employed to study the influence of geological and parametric uncertainty on solute transport. The proposed Markov model (see Chapter 3) is used to generate two-dimensional realizations of the geological structure. The methodology proposed in Chapter 5 is applied to investigate the parametric uncertainty inside each geological unit. Two types of hydraulic conductivity contrasts are investigated: low contrast and high contrast. The various flow scenarios are presented in Table (7.1). Table (7.2) shows the transition probabilities used to generate the geological structure. Fig.(7.1) shows one realization of the geological structure used in the experiments. The groundwater flow through the generated structure is introduced by a constant gradient between left and right boundary of the domain while, the top and bottom boundaries are assumed to be impervious. It can be seen, especially in case of high contrast, that the resulting flow pattern (computed equipotential lines and stream function) is strongly influenced by the spatial distribution of high and low conductivity zones.

7.3 Uncertainties Considered in The Experiments

The experiments address three types of uncertainties. Table (7.1) summarizes the scenarios suggested to study these uncertainties. The first set concerns geological uncertainty. In these experiments, it is assumed that the geological structure is uncertain, but the parameter value (hydraulic conductivity) inside each unit known with certainty (e.g. by sampling). The second set refers to geological and parametric uncertainty. In these experiments, it is assumed that both the geometry of the geological structures and the parameter value of individual unit

Table (7.1) Scenarios of The Numerical Experiments.

Contrast in Conductivity Uncertainty	Case Study 'Low Contrast'	Case Study 'High Contrast'
Geological Uncertainty	Set 1L	Set 1H
Geological + Parametric Uncertainty	Set 2L	Set 2H
Parametric Uncertainty	Set 3L	Set 3H

are considered to be subjected to uncertainty. In the third set parametric uncertainty is considered only. In these experiments, it is assumed that the geological structures and the borders between the different units are known deterministically (e.g. from surface and subsurface mapping, seismic data etc.), but the conductivities inside each lithology or rock are uncertain. This is a plausible geological assumption at some sites, like outcrops. The merit of this option is that it limits the ensemble of realizations to a subensemble of more likely realizations which are in agreement with the geological conditions.

7.4 Numerical Experiment Procedure

The procedure consists of the following steps:
Step 1: A geological structure is generated using the statistics displayed in Table (7.2). Fig.(7.1) shows a single realization of the structure.
Step 2: In case of geological uncertainty: a hydraulic conductivity value is assigned to each geological unit. However, in case of geological and parametric uncertainty or parametric uncertainty: a hydraulic conductivity field is generated inside each geological unit as described in Chapter 5. The statistical parameters used for generation of conductivity fields are presented in Table (7.3) and Table (7.4) for two case studies 'low contrast' and 'high contrast' respectively. One may notice that the values of the uncertainty in the conductivities, σ_K, are chosen 1/10th of the of the mean $\langle K \rangle$. This value will conserve the geological structure from being smeared by the parametric uncertainty. Fig.(7.2) shows the corresponding *PDF* of each case study. As one may notice, that each individual population is far apart from the others.
Step 3: The flow equation is solved with appropriate boundary conditions and the hydraulic head is computed. Fig.(7.1) shows the flownet of the single realizations presented of both case studies.

Table (7.2) Statistics of Fig.(7.1).

Length of The Given Section (m)=   300.
Depth of The Given Section (m)=   50.
Sampling interval in X-axis (m)=   2.
Sampling interval in Y-axis (m)=   1.

Input Statistics   Calculated Statistics

Horizontal Transition Probability Matrix

State	1	2	3	4	State	1	2	3	4
1	0.980	0.005	0.005	0.010	1	0.982	0.004	0.004	0.010
2	0.010	0.970	0.010	0.010	2	0.015	0.969	0.009	0.007
3	0.020	0.010	0.960	0.010	3	0.025	0.007	0.958	0.010
4	0.010	0.010	0.010	0.970	4	0.005	0.010	0.011	0.973

Vertical Transition Probability Matrix

State	1	2	3	4	State	1	2	3	4
1	0.600	0.100	0.200	0.100	1	0.760	0.071	0.123	0.046
2	0.100	0.700	0.100	0.100	2	0.187	0.606	0.054	0.154
3	0.100	0.100	0.700	0.100	3	0.174	0.086	0.606	0.135
4	0.100	0.100	0.100	0.700	4	0.132	0.053	0.059	0.756

Table (7.3) Parameters Used in Case of 'Low Contrast' in Conductivity.

Parameter State	w_i	$\langle K \rangle$ m/day	σ_K m/day	$\langle Y \rangle$	σ_Y	λ_x (m)	λ_y (m)
1	0.38	10.	1.0	2.30	0.1	5.	1.
2	0.17	8.	0.8	2.08	0.1	5.	1.
3	0.18	6.	0.6	1.79	0.1	5.	1.
4	0.27	4.	0.4	1.39	0.1	5.	1.

The parameters in the table: w_i is the marginal probability of state i, $\langle Y \rangle$ is the logarithmic transformation of $\langle K \rangle$, σ_Y is the logarithmic transformation of σ_K and λ_x, λ_y are the correlation lengths in x- and y-directions respectively.

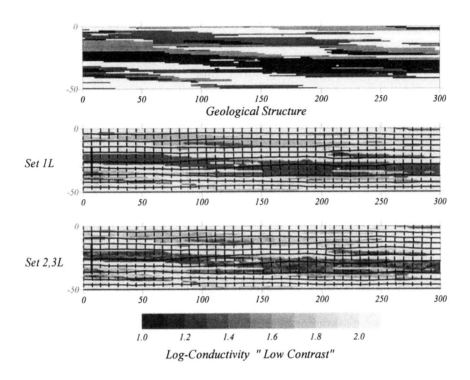

Geological Structure

Set 1L

Set 2,3L

Log-Conductivity " Low Contrast"

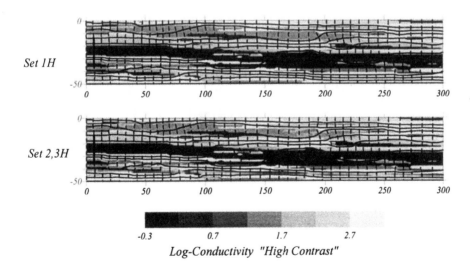

Set 1H

Set 2,3H

Log-Conductivity "High Contrast"

Fig.(7.1) Single Realizations of The Heterogeneous Structure with Flownet.

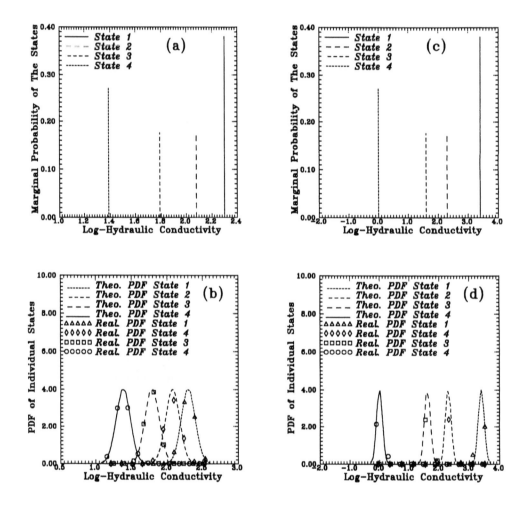

Fig.(7.2) Marginal Probabilities and PDFs of The States in The Experiments For Various Scenarios: (a) Set1L, (b) Set2L,3L, (c) Set1H, (d) Set2H,3H.

Table (7.4) Parameters Used in Case of 'High Contrast' in Conductivity.

Parameter State	w_i	$\langle K \rangle$ m/day	σ_K m/day	$\langle Y \rangle$	σ_Y	λ_x (m)	λ_y (m)
1	0.38	30.	3.0	3.4	0.1	5.	1.
2	0.17	10.	1.0	2.3	0.1	5.	1.
3	0.18	5.	0.5	1.6	0.1	5.	1.
4	0.27	1.	0.1	0.0	0.1	5.	1.

Step 4: The specific discharge is computed through the application of Darcy's law.

Step 5: A set of particles is released from a strip source located in the upstream part of the domain (see Fig.(7.1)). The initial condition is that no solute is present at time $t \leq 0$, $M_o = 100$ grams is injected via a strip with dimensions $W_o = 1$m in the main flow direction and $D_o = 20$m perpendicular to it. The centroid of the source is located at coordinates ($X_o = 5$m, $Y_o = -25$m) from the top left corner of the flow domain. Table (7.5) shows the physical and numerical parameters used in *MC* runs.

Step 6: The statistical characteristics of the concentration field are established by analyzing the multiple realizations and/or the spatial moments of the concentration fields. Estimation of the ensemble statistics are described below.

7.4.1 Ensemble Statistics of The Concentration Field

The mean concentration fields are constructed by averaging the solute concentrations over the total number of realizations (*MC=100*). Ensemble average concentrations can be determined at each grid point as,

$$\langle C(x,y,t) \rangle = \frac{1}{MC} \sum_{k=1}^{MC} C_k(x,y,t) \tag{7.1}$$

where, the angle brackets are shorthand for ensemble average (concentration expectation field), and $C_k(x,y,t)$ is the concentration at given t, location x,y in realization number k.

This ensemble average can be interpreted as an imaginary envelope of performance determined by all possible realizations [Fischer, 1975]. For a real field situation, however, there is always a single (actual) realization. Therefore, one could expect to find the actual plume within the imaginary envelope given by the expected value of the concentration field. The concentration variance at each grid point is given by,

$$\sigma_C^2 (x,y,t) = \frac{1}{MC} \sum_{k=1}^{MC} \left[C_k(x,y,t) - \langle C(x,y,t) \rangle \right]^2 \tag{7.2}$$

where, σ^2_C is the ensemble variance of the concentration which represent the uncertainty in the predictions.

The correct formula for Eq.(7.2) reads MC-1 in the denominator; MC is large and this factor is approximated; a common approach.

Table (7.5) Physical and Numerical Parameters Used in MC Experiments.

Parameter	Numerical Value
Effective Porosity	0.3
Longitudinal Micro-Dispersivity	0.1 m
Transverse Micro-Dispersivity	0.01 m
Injected Mass	M_o=100 grams
Strip Source Dimensions	W_o=1 m, D_o=20 m
Source Centroid Location	(X_o=5 m, Y_o=-25 m)
Number of Particles	N_p=3000 particles
Number of Time Steps	3000 steps
Domain Dimensions	L_x=300 m, L_y=50 m
Number of Grid Cells	N=7500 cells
Spatial Discretization	Δx =2 m, Δy =1 m
Up-Stream Head Boundary	1 m
Down-Stream Head Boundary	0 m
Time Step in Calculations	2 days
Number of Realizations/Set	MC=100
Hydraulic Conductivity	see Table (7.3) and Table (7.4)
Transition Probabilities	see Table (7.2)

7.4.2 Ensemble Statistics of The Plume Spatial Moments

In some situations, there may be no need to calculate the concentration distribution. It may be sufficient to delineate the affected zone due to a contaminant plume in aquifers. In such a case an estimation of the plume centroid and the spreading around it and their uncertainty would be adequate.

The ensemble average of the centroid displacement is calculated by,

$$\langle X_c(t) \rangle = \frac{1}{MC} \sum_{k=1}^{MC} X_{c_k}(t) \tag{7.3}$$

$$\langle Y_c(t) \rangle = \frac{1}{MC} \sum_{k=1}^{MC} Y_{c_k}(t) \tag{7.4}$$

and the ensemble standard deviation is computed as,

$$\sigma^2_{X_c}(t) = \frac{1}{MC} \sum_{k=1}^{MC} \left[X_{c_k}(t) - \langle X_c(t) \rangle \right]^2 \tag{7.5}$$

$$\sigma^2_{Y_c}(t) = \frac{1}{MC} \sum_{k=1}^{MC} \left[Y_{c_k}(t) - \langle Y_c(t) \rangle \right]^2 \tag{7.6}$$

The longitudinal ensemble variance of the particle cloud is calculated by,

$$\langle \sigma^2_{xx}(t) \rangle = \frac{1}{MC} \sum_{k=1}^{MC} \sigma^2_{xx_k}(t) \tag{7.7}$$

and the longitudinal ensemble variance of the variance of the particle cloud is calculated by,

$$Var\left(\sigma^2_{xx}(t) \right) = \frac{1}{MC} \sum_{k=1}^{MC} \left[\sigma^2_{xx_k}(t) - \langle \sigma^2_{xx}(t) \rangle \right]^2 \tag{7.8}$$

7.4.3 Ensemble Statistics of The Longitudinal Macro-Dispersion Coefficient

The ensemble longitudinal macro-dispersion coefficient is calculated by,

$$\langle D_{xx}(t) \rangle = \frac{1}{MC} \sum_{k=1}^{MC} D_{xx_k}(t) \tag{7.9}$$

and the ensemble variance of the macro-dispersion coefficient is calculated as follows,

$$Var\left(D_{xx}(t)\right) = \frac{1}{MC} \sum_{k=1}^{MC} \left[D_{xx_k}(t) - \langle D_{xx}(t) \rangle \right]^2 \tag{7.10}$$

7.5 Graphical Representation of Monte-Carlo Experimental Results

The results of the *MC* experiments are presented in a series of figures and graphs. Table (7.6) summarizes the figure captions of the snapshots of the plumes and their statistics in each case study.

Ensemble concentration profiles and standard deviation profiles are presented in Fig.(7.12) for the case of low permeability contrast and in Fig.(7.24) in case of high contrast. The spatial moments of ensemble plumes with some single realizations are also presented. Only 3 out of 100 realizations are displayed for demonstration purposes. Fig.(7.13) and Fig.(7.14) show the spatial moments in case of low contrast. Fig.(7.25) and Fig.(7.26) show the spatial moments in case of high contrast.

Table (7.6) Summary of Figure Captions of The Experiments.

Fig.	Figure Captions		
Set of Exp.	Single Real., C,	$\langle C \rangle$	σ_C
Set 1L	Fig.(7.3)	Fig.(7.4)	Fig.(7.5)
Set 2L	Fig.(7.6)	Fig.(7.7)	Fig.(7.8)
Set 3L	Fig.(7.9)	Fig.(7.10)	Fig.(7.11)
Set 1H	Fig.(7.15)	Fig.(7.16)	Fig.(7.17)
Set 2H	Fig.(7.18)	Fig.(7.19)	Fig.(7.20)
Set 3H	Fig.(7.21)	Fig.(7.22)	Fig.(7.23)

7.6 Discussion of The Results

7.6.1 Case of Low Contrast in Hydraulic Conductivity

7.6.1.1 Geological Uncertainty

Snapshots of the plume in a single realization are presented in Fig.(7.3). The corresponding conductivity field and flow pattern of this typical realization are displayed in Fig.(7.1) set 1L. These snapshots show the plume at 200, 1000, 1600, 2000, 3000 days, respectively. One would expect that the behaviour of each realization to be different. Fig.(7.4) shows the evolution of the ensemble plumes at the same snapshots. The ensemble plumes are smooth in appearance. In the single realization the shape of the structure is manifested in the plume evolution. Some parts of the plume are moving faster than the others due to the contrast in conductivity between the adjacent units.

It can be noticed that the plume is covering only certain part of the whole formation. This part is in the same order of magnitude of the maximum plume size in a single realization. The ensemble standard deviation of the plumes is shown in Fig.(7.5). The ensemble standard deviation is larger than the ensemble mean. This means that the concentration prediction is subjected to a relatively high degree of uncertainty. The magnitude of the ensemble concentration mean and standard deviation can be seen more clearly from the profiles in Fig.(7.12).

7.6.1.2 Geological and Parametric Uncertainty

Snapshots of the plume in a single realization are presented in Fig.(7.6). The corresponding conductivity field and flow pattern of this typical realization is displayed in Fig.(7.1) set 2L. These snapshots show the plume at 200, 1000, 1600, 2000, 3000 days, respectively. As in the previous case, one would expect that the behaviour of each realization to be different. Fig.(7.7) shows the evolution of the ensemble plumes at the same snapshots. The ensemble plumes is smooth in appearance. In a single realization the shape of the structure is manifested in the plume evolution. The parametric uncertainty does not show remarkable difference in the plume shape in comparison with the previous case. This has to do with the degree of uncertainty considered in each geological unit which is relatively low. This is also manifested in the flownet shown in Fig.(7.1) set 1L and set 2L where one cannot observe a remarkable differences in the flow pattern.

Similar results as shown in the previous case can be reported. It can be noticed that the solute plume is covering only certain part of the whole formation. This part is in the same order of magnitude of the maximum plume size in a single realization. The ensemble standard deviation of the plumes are shown in Fig.(7.8). The standard deviation is larger than the ensemble mean. This means that the concentration prediction is subjected to a relatively high degree of uncertainty. The magnitude of the ensemble concentration mean and standard deviation can be seen in Fig.(7.12). The uncertainty in this case is slightly higher than in the previous case because the magnitude of parametric uncertainty is relatively small.

7.6.1.3 Parametric Uncertainty

Snapshots of the plume in a single realization are presented in Fig.(7.9). The corresponding conductivity field and flow pattern of this typical realization are displayed in Fig.(7.1) set 3L. This realization is the same as the previous single realization shown in Fig.(7.6). These snapshots show the plume at 200, 1000, 1600, 2000, 3000 days respectively. In this case, the plume behaviour in each realization does not vary so much from one another. The reason is that all the plumes of different realizations are controlled by the same geological pattern, in other words the plume behaviour is conditioned by a given geological structure. Therefore, the behaviour in single realization, Fig.(7.9), is very close to the ensemble plumes shown in Fig.(7.10).

It can be seen that the solute plume is covering only certain portion of the whole formation. This part is in the same order of magnitude of the maximum plume size in a single realization. The ensemble standard deviation of the plumes are shown in Fig.(7.11). The shape of plume evolution in single realization, Fig.(7.9), the ensemble plumes, Fig.(7.10), and the ensemble standard deviation in concentration, Fig.(7.11), display the same pattern. The standard deviation is less than the ensemble mean in this case. This means that the uncertainty in concentration field is relatively low in comparison with the previous two cases. The magnitude of the ensemble concentration mean and standard deviation can be seen from the profiles in Fig.(7.12).

Close observation of Fig.(7.12) shows that the uncertainty decreases with time. The ensemble standard deviation in concentration in early times (time=200 days) is high near the source. However at later times (time=2000 days) it decreases significantly. The reason for that is the plume size and the spatial arrangements of the heterogeneous units. Near the source the plume is small and it has a large degree of freedom to spread in different forms from one realization to another according to the spatial arrangements of the geological units in each realization. This causes relatively large uncertainty near the source. However, far from the source the plume is becoming larger and it covers large area of the aquifer. Then the degree of freedom to spread in one realization to another is not that high and the uncertainty is less. This result is qualitatively in the same line with Gelhar [1993]. Gelhar proved analytically that concentration uncertainty is proportional to the concentration gradient. Near the source the concentration gradient is high and consequently the uncertainty is high. However, far from the source the concentration gradient is low and consequently low uncertainty is expected.

7.6.1.4 Analyzing Plume Spatial Moments

In case of geological and both geological and parametric uncertainty: the results of three single realizations are displayed in Fig.(7.13). It can be observed that each realization shows a different plume velocity and a different spreading pattern. The three different realizations show considerable scatter in the longitudinal variance. The 95% confidence intervals produce a wide envelope over the ensemble mean in case of geological uncertainty. This envelope is becoming wider in case of geological and parametric uncertainty. One may recognize that one of the realization is above the envelope of 95% confidence interval in both cases. This might happen because with 95% interval there still a chance of falling outside the envelop. However, in case of parametric uncertainty only the 95% confidence intervals are very narrow (see Fig.(7.14)). All the realizations are falling within the envelope.

The ensemble longitudinal macro-dispersion coefficient in case of geological and both geological and parametric uncertainty, Fig.(7.13) show a monotonous increase with time, while in each realization it fluctuates. The ensemble macro-dispersion coefficient emphasizes that the macro-dispersion coefficient is different from the pore scale dispersion coefficient. A closer look shows that the magnitude of the macro-dispersion coefficient in case of geological and parametric uncertainty ($<D_{xx}>=0.27$ m^2/day) is slightly higher than in the case of geological uncertainty ($<D_{xx}>=0.25$ m^2/day). In case of parametric uncertainty only, the behaviour of the ensemble macro-dispersion coefficient (Fig.(7.14)) shows up and down growth due to the influence of the geological structure. In this case an asymptotic macro-dispersion cannot be defined.

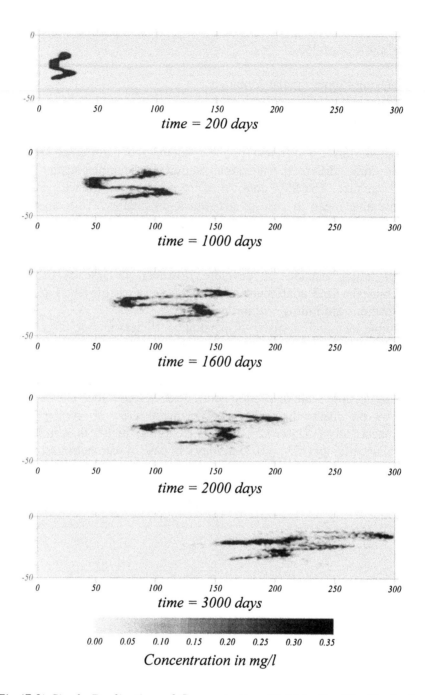

Fig.(7.3) Single Realization of Concentration Field in Set 1L: Geological Uncertainty with Low Contrast.

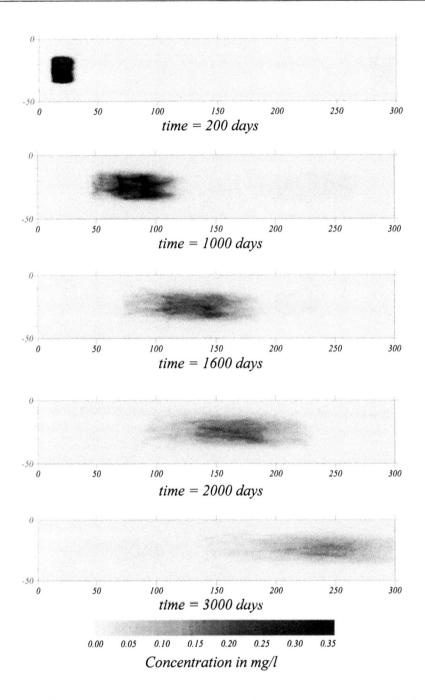

Fig.(7.4) Ensemble Average Concentration Field in Set 1L: Geological Uncertainty with Low Contrast.

219

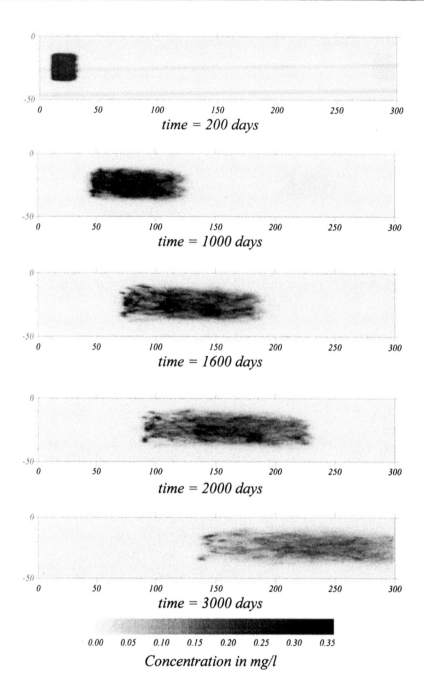

Fig.(7.5) Ensemble Standard Deviation in Concentration Field in Set 1L: Geological Uncertainty with Low Contrast.

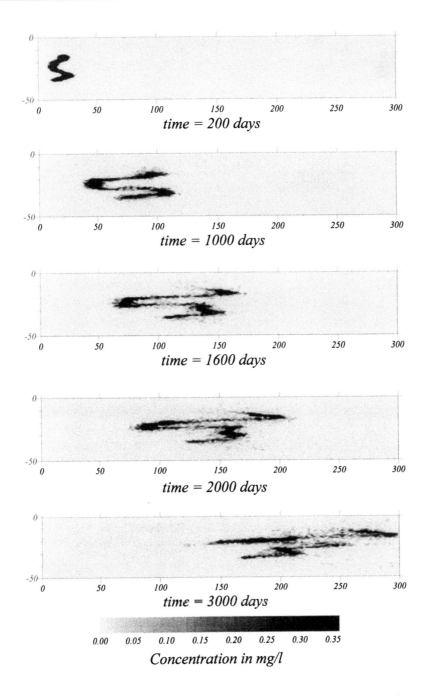

time = 200 days

time = 1000 days

time = 1600 days

time = 2000 days

time = 3000 days

0.00   0.05   0.10   0.15   0.20   0.25   0.30   0.35

Concentration in mg/l

Fig.(7.6) Single Realization of Concentration Field in Set 2L: Geological and Parametric Uncertainty with Low Contrast.

221

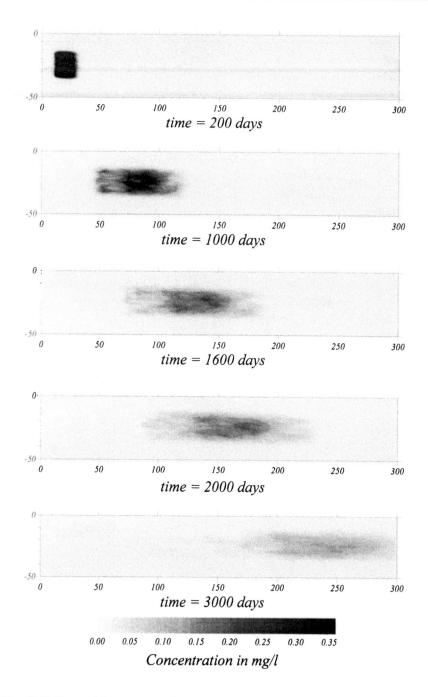

Fig.(7.7) Ensemble Average Concentration Field in Set 2L: Geological and Parametric Uncertainty with Low Contrast.

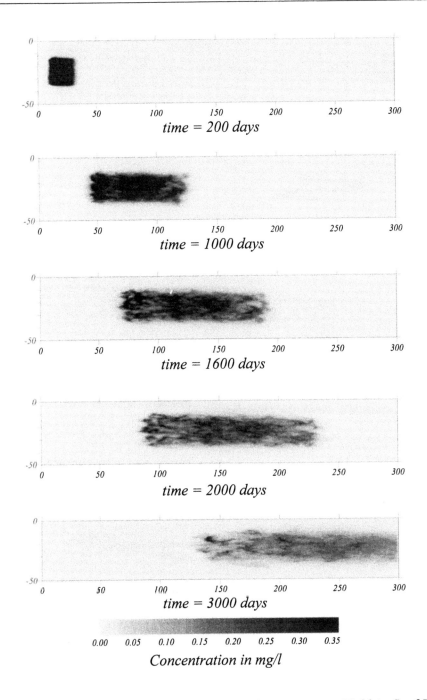

time = 200 days

time = 1000 days

time = 1600 days

time = 2000 days

time = 3000 days

0.00   0.05   0.10   0.15   0.20   0.25   0.30   0.35

Concentration in mg/l

Fig.(7.8) Ensemble Standard Deviation in Concentration Field in Set 2L: Geological and Parametric Uncertainty with Low Contrast.

223

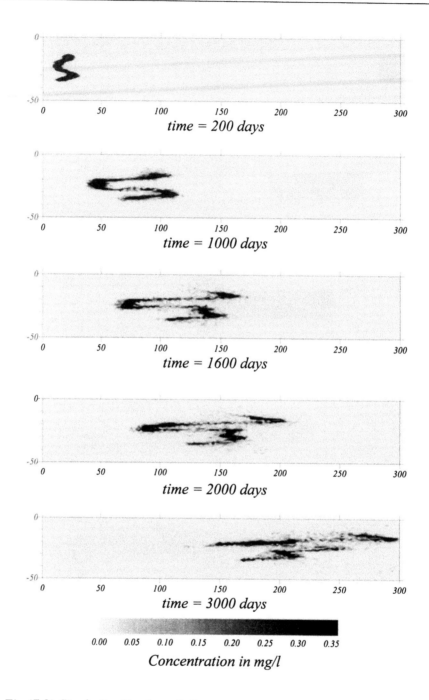

Fig.(7.9) Single Realization of Concentration Field in Set 3L: Parametric Uncertainty with Low Contrast.

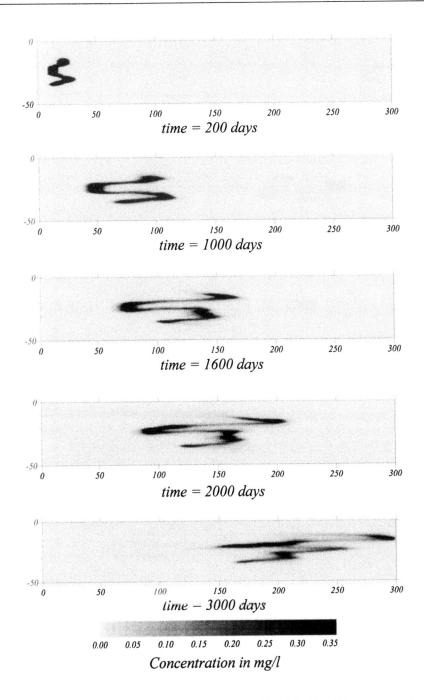

time = 200 days

time = 1000 days

time = 1600 days

time = 2000 days

time − 3000 days

0.00   0.05   0.10   0.15   0.20   0.25   0.30   0.35

Concentration in mg/l

Fig.(7.10) Ensemble Average Concentration Field in Set 3L: Parametric Uncertainty with Low Contrast.

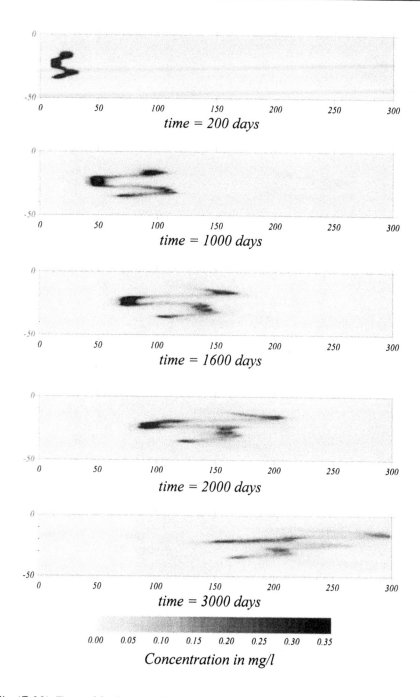

Fig.(7.11) Ensemble Standard Deviation in Concentration Field in Set 3L: Parametric Uncertainty with Low Contrast.

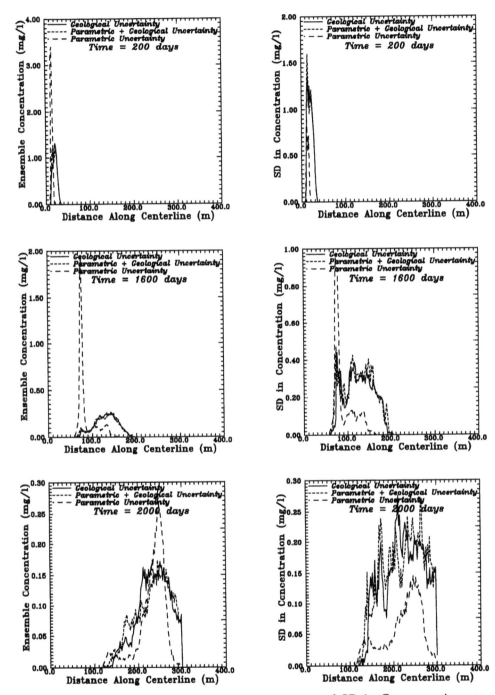

Fig.(7.12) Profiles of Ensemble Concentration and SD in Concentration at Some Selected Snapshots (Low Contrast).

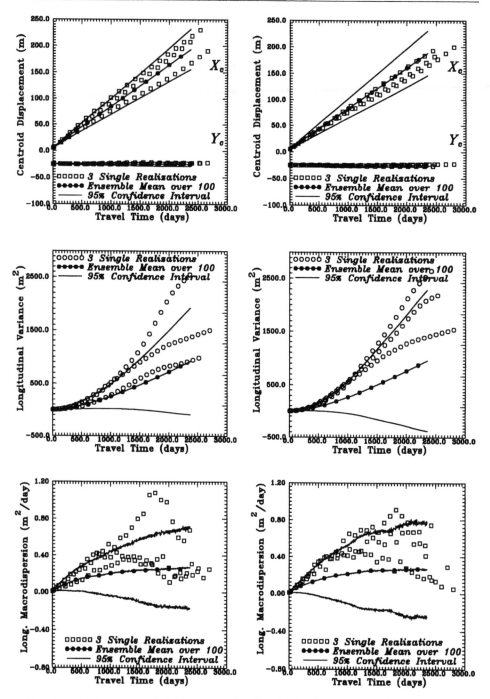

Fig.(7.13) Plume Spatial Moments for Set 1,2L: Geological (left) and Geological+Parametric (right) Uncertainty with Low Contrast.

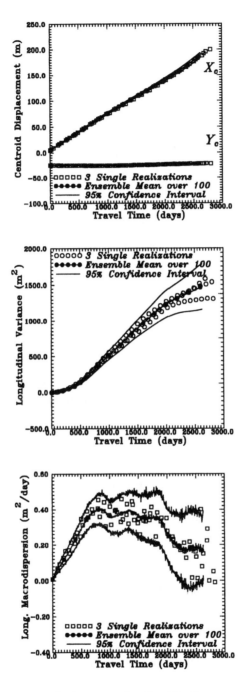

Fig.(7.14) Plume Spatial Moments for Set 3L: Parametric Uncertainty with Low Contrast.

7.6.2 Case of High Contrast in Hydraulic Conductivity

7.6.2.1 Geological Uncertainty

Snapshots of the plume in a single realization are presented in Fig.(7.15). The corresponding conductivity field and flow pattern of this realization are displayed in Fig.(7.1) set 1H. These snapshots show the plume at 200, 1000, 1600, 2000, 3000 days, respectively. One would expect that the behaviour of each realization is different from the previous case study. Fig.(7.16) shows the evolution of the ensemble plumes at the same snapshots. The ensemble plumes are smooth in appearance. However, in single realization the shape of the structure is manifest in the plume evolution and the concentration is sharp in appearance. Some parts of the plume are moving faster than others due to the contrast in conductivity between the adjacent units which is higher in this case than in the pervious case.

It can be noticed that the solute plume is covering almost the whole formation. This is due to the existence of relatively low permeable zones where the particles are trapped in these zones and take long time to be released. However, the particles in the high permeable zones are travelling significantly faster and leaving the domain. The size of the ensemble plume is in the same order of magnitude of the maximum plume size in a single realization. The ensemble standard deviation of the plumes is shown in Fig.(7.17). As shown in the previous case the standard deviation is larger than the ensemble mean which means that the concentration prediction is subjected to a relatively high degree of uncertainty. The magnitude of the ensemble concentration mean and standard deviation can be seen more clearly from the concentration profiles in Fig.(7.24).

7.6.2.2 Geological and Parametric Uncertainty

Snapshots of the plume in a single realization are presented in Fig.(7.18). The corresponding conductivity field and flow pattern of this realization are displayed in Fig.(7.1) set 2H. These snapshots show the plume at 200, 1000, 1600, 2000, 3000 days, respectively. Fig.(7.19) shows the evolution of the ensemble plumes at the same snapshots. The ensemble plumes are smooth in appearance. However, in a single realization the shape of the structure is manifested in the plume evolution. The parametric uncertainty does not show remarkable difference in the plume shape in comparison with geological uncertainty only. As mentioned earlier, this has to do with the degree of uncertainty considered in each geological unit which is relatively low. This is also manifest in the flownet shown in Fig.(7.1) set 1H and set 2H.

Similar results as in the case of geological uncertainty only can be reported. It can be noticed that the solute plume is covering almost the whole formation.

The size of the ensemble plume is in the same order of magnitude of the maximum plume size in a single realization. The ensemble standard deviation of the plumes are shown in Fig.(7.20). The standard deviation is larger than the ensemble mean. That means, the concentration prediction is subjected to a relatively high degree of uncertainty. The magnitude of the ensemble mean concentration and standard deviation along the longitudinal centre line of the plume can be seen more clearly in profiles of Fig.(7.24). The uncertainty in this case is slightly higher than the previous case. One may realize that the uncertainty in this case spans the whole formation considered.

7.6.2.3 Parametric Uncertainty

Snapshots of the plume in a single realization are presented in Fig.(7.21). The corresponding conductivity field and flow pattern of this realization is displayed in Fig.(7.1) set 3H. This realization is the same as the previous single realization. These snapshots show the plume at 200, 1000, 1600, 2000, 3000 days respectively. In this case, the plume behaviour in each realization does not vary so much from one another, but it is smoother. The reason is that all the plumes of different realizations are controlled by the same geological pattern, in other words conditioned by the geological structure. Therefore, the behaviour in single realization, Fig.(7.21), is very close to the ensemble plumes shown in Fig.(7.22).

It can be seen that the solute plume is covering almost the whole formation. The size of the ensemble plumes has the magnitude of the maximum plume size in a single realization. The ensemble standard deviation of the plumes are shown in Fig.(7.23). The shape of plume evolution in a single realization (see Fig.(7.21)), the ensemble plumes (Fig.(7.22)), and the standard deviation in concentration (Fig.(7.23)) display the same pattern. The standard deviation is less than the ensemble mean in this case. This means that the uncertainty in concentration field is relatively low. The magnitude of the ensemble mean concentration and the standard deviation along the longitudinal centre line of the plume can be seen in Fig.(7.24). The uncertainty in this case is low in comparison with the previous two cases.

Close observation of Fig.(7.24) shows, as in case of low contrast, that the uncertainty decreases with time. The ensemble standard deviation in concentration at early times (time=200 days) is high. However at later times (time=2000 days) is less. The reason for that has been mentioned before (section 7.6.1.3). The only difference here is that the ensemble variance is distributed over the whole formation with high values near the source and low values near the plume face, far from the source.

7.6.2.4 Analyzing Plume Spatial Moments

In case of geological and both geological and parametric uncertainties, the results of three single realizations are displayed in Fig.(7.25). It can be observed that each realization show different plume velocity and different spreading pattern as in the low conductivity contrast. However, the uncertainty here is higher. The three different realizations show considerable scatter in both the centroid displacement and the longitudinal variance. The 95% confidence intervals produce wider envelope over the ensemble mean than in the case of low contrast.

The envelope of uncertainty of the plume centroid in case of geological and parametric uncertainty is less wide than in case of geological uncertainty only (see Fig.(7.25)). This is in contrary to the case of low contrast, but on the other hand the centroid is retarded in case of geological and parametric uncertainty. This reflects the influence of the log-normal distribution in conductivity which does not appear in case of low contrast because at small variances the normal and log-normal distribution are very close. When the log-normality is dominated, then the values of permeability less than the arithmetic mean is more likely to occur than the values of permeability more than the arithmetic mean. Therefore, the geological units are dominated by zones of permeability less than the arithmetic mean. This causes the retardation in the centroid movement and a retardation in the development of its uncertainty.

One may recognize that all realizations are falling within the envelope of 95% confidence interval in both cases. This is because of the wide envelope. However, in case of parametric uncertainty only the 95% confidence intervals are very narrow (see Fig.(7.26)).

The ensemble longitudinal macro-dispersion coefficient in case of geological and both geological and parametric uncertainties, Fig.(7.25), shows a monotonous growth with time, while in each realization it fluctuates.

The ensemble asymptotic cannot be reached in the time of the experiment. In case of parametric uncertainty, the behaviour of the ensemble macro-dispersion coefficient shows up and down growth due to the influence of the geological structure.

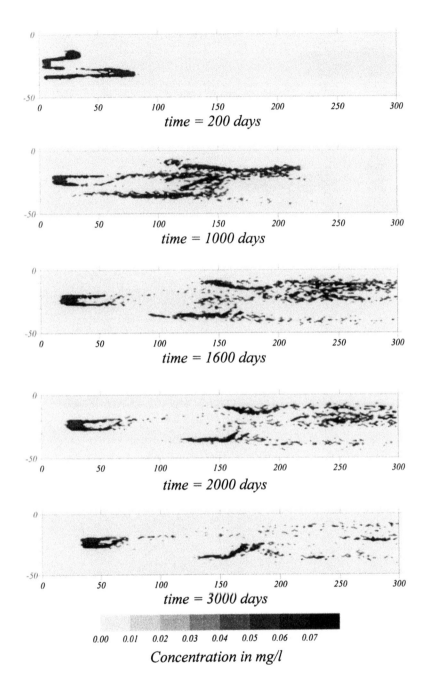

time = 200 days

time = 1000 days

time = 1600 days

time = 2000 days

time = 3000 days

0.00   0.01   0.02   0.03   0.04   0.05   0.06   0.07

Concentration in mg/l

Fig.(7.15) Single Realization of Concentration Field in Set 1H: Geological Uncertainty with High Contrast.

233

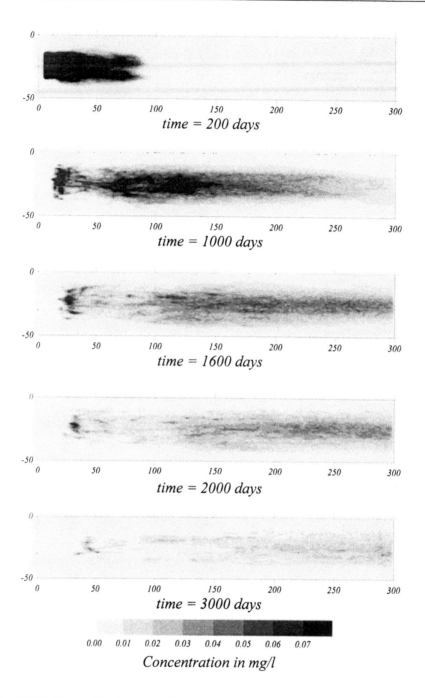

Fig.(7.16) Ensemble Average Concentration Field in Set 1H: Geological Uncertainty with High Contrast.

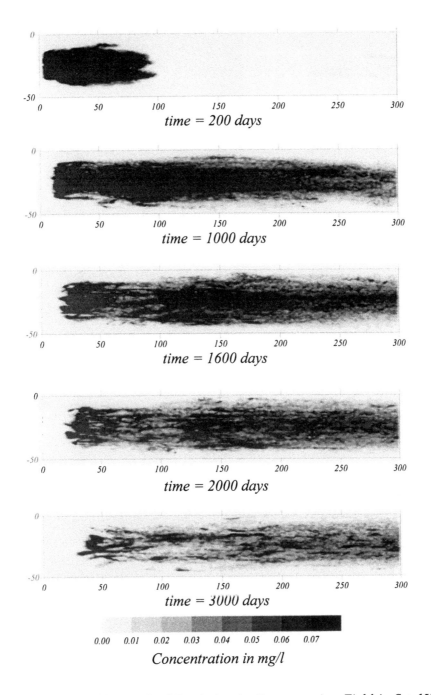

time = 200 days

time = 1000 days

time = 1600 days

time = 2000 days

time = 3000 days

0.00   0.01   0.02   0.03   0.04   0.05   0.06   0.07

Concentration in mg/l

Fig.(7.17) Ensemble Standard Deviation in Concentration Field in Set 1H: Geological Uncertainty with High Contrast.

235

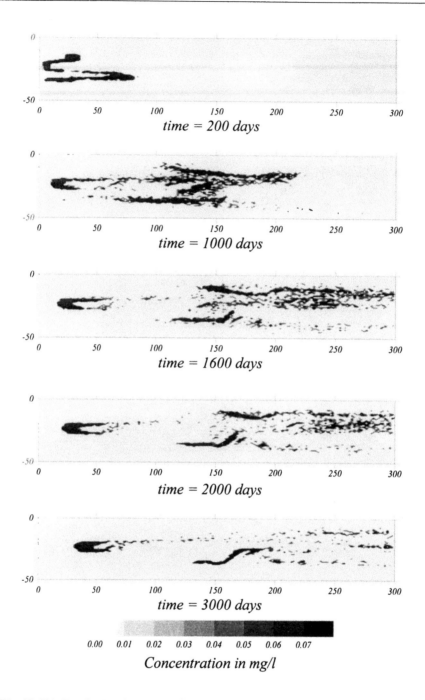

Fig.(7.18) Single Realization of Concentration Field in Set 2H: Geological and Parametric Uncertainty with High Contrast.

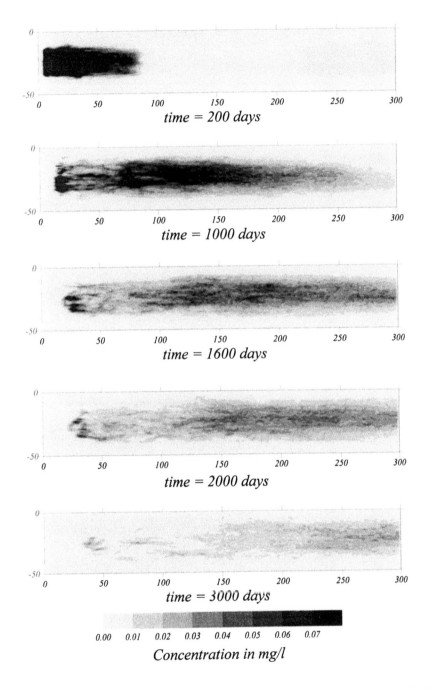

Fig.(7.19) Ensemble Average Concentration Field in Set 2H: Geological and Parametric Uncertainty with High Contrast.

237

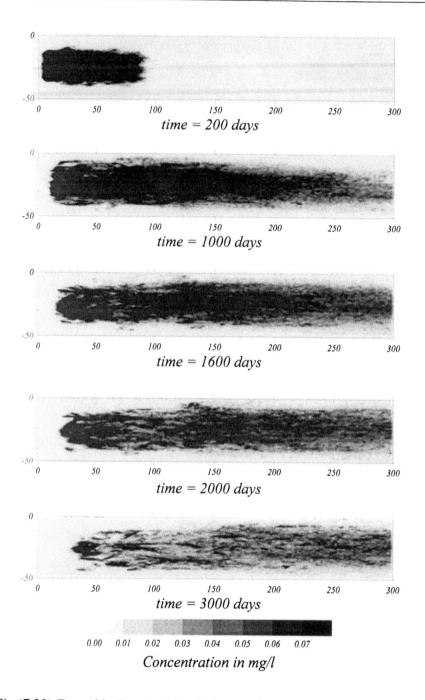

Fig.(7.20) Ensemble Standard Deviation in Concentration Field in Set 2H:
Geological and Parametric Uncertainty with High Contrast.

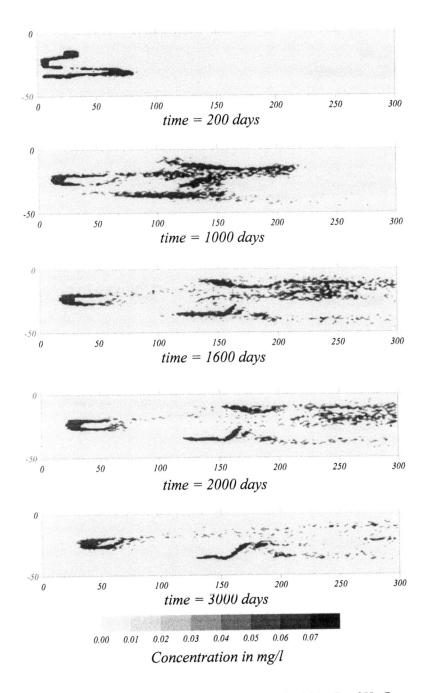

Fig.(7.21) Single Realization of Concentration Field in Set 3H: Parametric Uncertainty with High Contrast.

239

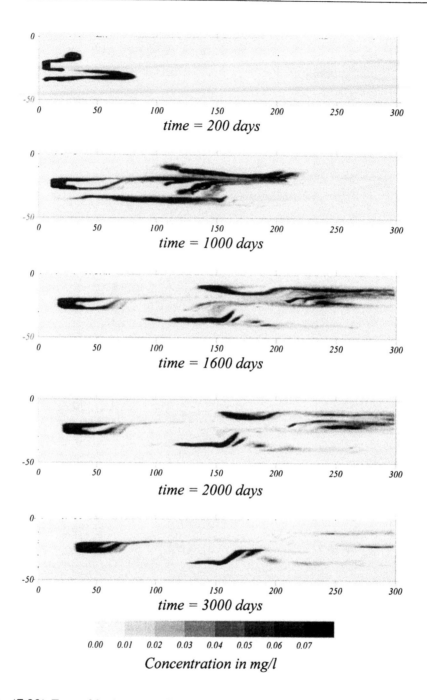

Fig.(7.22) Ensemble Average Concentration Field in Set 3H: Parametric Uncertainty with High Contrast.

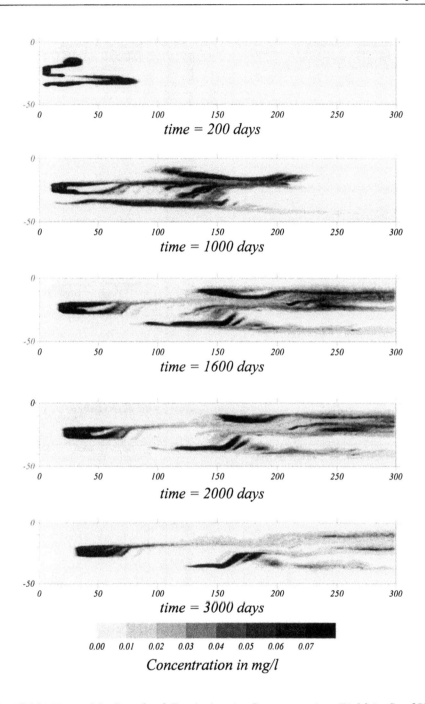

time = 200 days

time = 1000 days

time = 1600 days

time = 2000 days

time = 3000 days

0.00 0.01 0.02 0.03 0.04 0.05 0.06 0.07

Concentration in mg/l

*Fig.(7.23) Ensemble Standard Deviation in Concentration Field in Set 3H:
Parametric Uncertainty with High Contrast.*

241

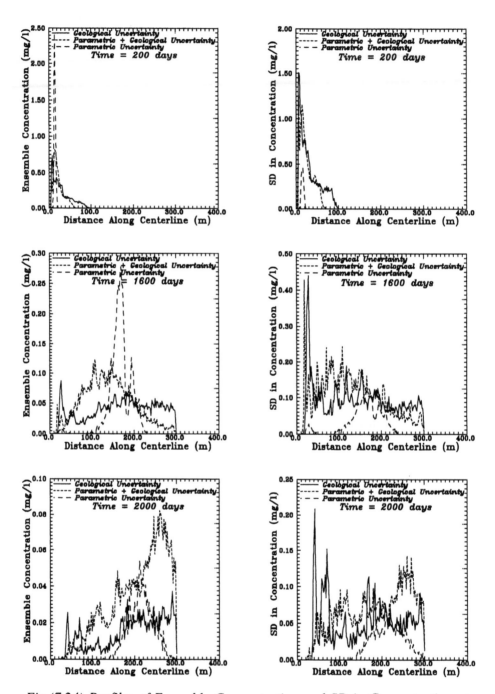

Fig.(7.24) Profiles of Ensemble Concentration and SD in Concentration at Some Selected Snapshots (High Contrast).

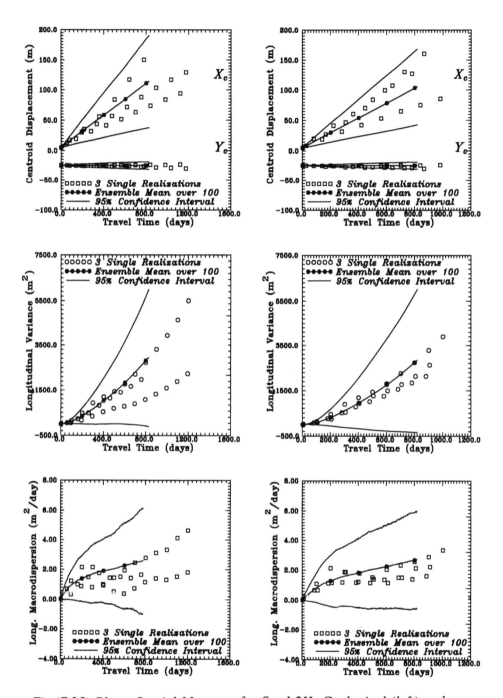

Fig.(7.25) Plume Spatial Moments for Set 1,2H: Geological (left) and Geological+Parametric (right) Uncertainty with High Contrast.

243

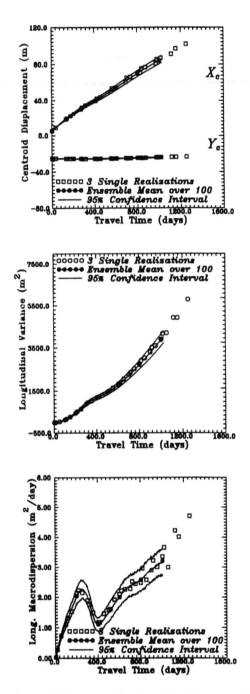

Fig.(7.26) Plume Spatial Moments for Set 3H: Parametric Uncertainty with High Contrast.

7.7 Summary and Conclusions

The irregular shapes of solute plumes in aquifers, due to the heterogeneity of the natural formations and the inability of characterizing these heterogeneities in a deterministic sense, suggest that it may be more realistic to make predictions in a probabilistic context rather than in the traditional deterministic framework.

In this chapter an evaluation of the uncertainty in predicting advective-dispersive solute transport is presented. Three different kinds of uncertainties are considered: parametric, geological, and combined uncertainties.

A multi-realizations approach (*MC* approach) has been followed to estimate the uncertainty associated with solute concentration (for assessment of the concentration at specific location in space and time) and/or plume spatial moments (for delineation of the contaminated zone covered by the plume). The results are based on the assumption that the geological variability can be characterized by the coupled chain Markov model (Chapter 3) and log-normal distribution of hydraulic conductivity with exponential auto-correlation structure inside each geological unit (Chapter 5). The uncertainty in the system description is characterized by calculating the ensemble variance over many (100) equally probable realizations of the system generated by the methodology suggested in this study. Next, some concluding remarks are given:

(1) The *MC*-approach presented here demonstrates the potential applicability of the coupled Markov chain model in quantifying geological uncertainty when the geological system possesses a Markovian property and discrete features are present in the system. This idea has been supported by some field observations [Krumbein, 1967].

(2) Lack of information about both the geological setting and parametric values of each geological unit cause the greatest uncertainty in the predictions. This uncertainty can be strongly reduced if sufficient information about the geological system is known in a deterministic sense.

(3) For the multi-realization experiments, it is emphasized that the ensemble macro-dispersion coefficients are not equivalent to those used for pore scale dispersion process (microscopic dispersion). The same has been found for the individual macro-dispersion coefficient in single realization experiments (see Chapter 6).

(4) The evolution of the plume spatial moments in a single realization or the ensemble spatial moments in case of parametric uncertainty do not display a monotonous growth with time. However, the ensemble spatial moments in case of geological, and both geological and parametric uncertainty show clearly a

monotonous growth with time although only 100 *MC* runs are used in the experiments.

(5) In case of geological uncertainty and combined parametric and geological uncertainty, the concentration fields and the spatial moments of actual plume (single realization) can differ significantly from their expected values (average over ensemble). However, in case of parametric uncertainty only, the behaviour of single plume is similar to the behaviour of ensemble plumes. This is due to the influence of conditioning of the generated realizations by a subensemble of realizations that possess the same geological structure.

(6) The imaginary envelope of the ensemble plumes in case of relatively high contrast is wider than the imaginary envelope of the ensemble plumes in case of relatively low contrast.

(7) The various types of uncertainty show a decrease with time from early times near the source to later times far from the source. This behaviour is, qualitatively, in agreement with Gelhar [1993].

(8) In case of relatively low contrast in conductivity: there is no remarkable difference on plume spatial moments and the evolution of their uncertainty under geological or geological and parametric uncertainty. However, in case of relatively high contrast in conductivity: it has been demonstrated that there is a retardation of the plume spatial moments and a retardation in the evolution of their uncertainty for combined geological and parametric uncertainty. This is due to the influence of the lognormality of the conductivity which is more pronounced in case of relatively high contrast.

(9) A fixed geological structure (a single realization) gives a significantly different dispersion behaviour from the ensemble mean over many realizations (100). It has been found that the contaminated area can be indicated with certainty, however, the local concentration at fixed point within the area is subjected to high uncertainty. Outside the contaminated area the concentration is practically zero, with certainty. This condition marks the practical value of the approach developed in this study.

CHAPTER 8

Laboratory Experiments on Solute Transport in Homogeneous and Heterogeneous Media

8.1 Introduction

For the simulation of groundwater contaminant transport numerical models are the most appropriate tools, due to their great flexibility and relatively low costs. However, validation of the concepts upon which these models are based, experimental data from field studies or laboratory tests are needed. For the support of new modelling approaches, field data are scarce and in general not sufficiently detailed. Therefore, a physical model has been built for studying groundwater solute transport in artificial homogeneous and heterogeneous aquifer. This model provides us with laboratory controlled conditions for the experimental tests that lead to a better understanding of transport in heterogeneous media.

The motivation of this experimental work is to get insight into the influence of heterogeneity on the phenomenon of dispersion and to distinguish the various dispersion regimes.

Experimental studies of solute movements by groundwater in saturated homogeneous soils have been carried out since Bear [1961a]. However, transport phenomenon in heterogeneous soils has not been given much attention in two-dimensional experimental models. The results of the laboratory model are compared with results predicted using numerical models developed in this study.

8.2 Pilot Test

A small pilot study has been carried out at Geotechnical Laboratory of the Technical University of Delft. The equipment consists of a box of dimensions $60 \times 43 \times 1$ cm^3. With the information and experience gained from the pilot study a final set-up has been constructed. The set-up is described in the next section.

8.3 The Experimental Set up

This apparatus consists of a box model constructed of 1.2 cm thick Perspex sheets and filled with unconsolidated glass beads as shown in Fig.(8.1). The sandbox is considered to represent a vertical, two-dimensional cross section of a confined aquifer in the longitudinal direction, with a limited third dimension. The inside dimensions of the working section of the box are 100 cm long, 56 cm wide, and 1 cm deep [see Fig.(8.1)]. The ends of the box consist of constant head tanks which are separated from the glass beads chamber by glass fibber filters to keep the glass beads out of the head tanks. Influent and effluent reservoirs are connected to the constant head tanks by tubes. A pump is used at the influent tank for circulating the water in the system. The pump is capable of delivering a maximum of 6 l/min. By manipulating the water level in the inflow and outflow tanks different horizontal discharge rates and Darcy's fluxes can be established and a uniform groundwater flow field is obtained.

8.4 Porous Media and Method of Packing

Different-sized pure silica glass beads are selected and used to model the heterogeneity. Fig.(8.2) shows the glass beads used in the experiments. Table (8.1) shows the size range and the average diameter, d_m, of the glass beads. The box is filled with one size of beads, in case of a homogeneous sample. It is filled with different sized beads in case of heterogeneous samples. The filling and packing technique is as follows. Before packing, the beads are soaked in distilled water, washed and dried in 110°C oven to remove any impurities due to manufacture. The model is filled partially with distilled water. Then, the beads are filled from the top into the model by a funnel. During the filling process the model walls are blown by a hammer. Packing is guided by predesigned arrangement of the media, to produce a heterogeneous porous medium with appropriate configuration. The porous medium rests directly against the bottom and sides of the model. From the top a strip of sponge rubber confines the top. The sponge rubber prevents any flow short-circuiting along the top of the medium. In case of heterogeneous samples, some filter rules should be followed to avoid segregation at the interfaces (the small sized grains fill the pores of the big sized ones).

Some segregation, however, cannot be avoided during packing, especially for the very fine glass beads. Three packings are selected and used for presentation here. The first is a homogeneous, the second contains block-shaped heterogeneity (medium 1) and the third is packed in the form of perfect and imperfect horizontal layers (medium 2).

(a)

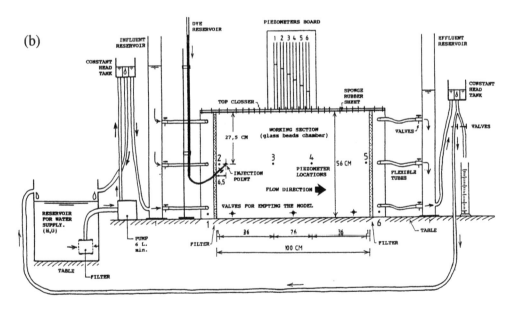

(b)

Fig.(8.1) Laboratory Set-Up: (a) Photogarph, (b) Sketch.

Table (8.1) Physical Parameters of The Glass Beads.

Beads Size (mm)	d_m (mm)	ε_t	Measured K (cm/sec)		Empirical K (cm/sec)		α_l (mm)
			Loose	Dense	Hazen	K-C*	
.08 -.11	0.095	48%	.0064	.0046	.009	.008	0.018
.4 -.52	0.46	48%	.314	.125	.212	.186	0.353
.85 -1.23	1.04	48%	1.09	0.66	1.08	.953	0.800
2.±.5	2.0	48%	2.08	1.93	4.0	3.52	1.405

* Kozeny-Carman

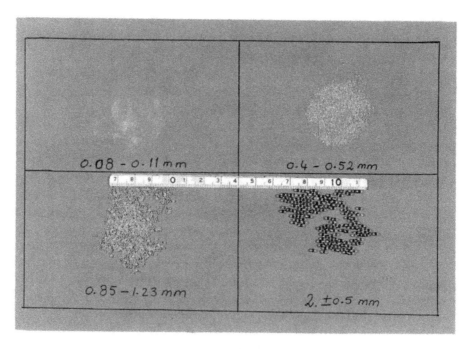

Fig.(8.2) Glass Beads Used in The Experiments.

8.5 Parameters Estimation of Media Properties

8.5.1 Estimation of Porosity

8.5.1.1 Total Porosity

The total porosity, ε_t, is defined as the ratio of pore volume to total volume,

$$\varepsilon_t = \frac{V_v}{V_t} \tag{8.1}$$

where, V_v is the volume of voids and V_t is the total volume of the sample.

The total porosity for the glass beads is calculated as follows. The volume of water occupying the pore spaces is determined by weighting a cup with beads when dry and again after it is saturated,

$$V_v = \frac{W_{sat} - W_{dry}}{\gamma_w} \tag{8.2}$$

where, γ_w is the specific weight of water, W_{sat} the saturated weight of the sample and W_{dry} is the dry weight of the sample.

The estimated total porosity are tabulated in Table (8.1). The 48% indicates loose packing.

8.5.1.2 Effective Porosity

The porosity values used in the transport equation are the effective porosity, ε, which is slightly less than the total porosity. It is required for the calculation of the real velocity of the flow, V. The real velocity is related to the specific discharge, q, by the relation,

$$V = \frac{q}{\varepsilon} \tag{8.3}$$

From a preliminary tracer test, the effective porosity of beads ($d_m=0.46$ mm) is estimated. The plume centroid is recorded at regular intervals and plotted in Fig.(8.3). A best fit line is plotted to the experimental data which led to a slope corresponding to the real velocity,

$$V = \frac{d\langle X \rangle}{dt} \qquad (8.4)$$

where, $\langle X \rangle$ is the centroid displacement of the plume from the injection point and t is the corresponding time.

The specific discharge is calculated by the volume of water collected within a certain period of time, Q, divided by the model cross-sectional area,

$$q = \frac{Q}{B\,D} \qquad (8.5)$$

where, B is the model depth (56 cm) and D is the model thickness perpendicular to the flow direction (1 cm). From this experiment the effective porosity of (d_m=0.46 mm) is estimated to be about 40%.

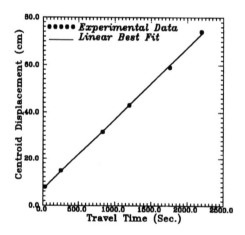

Fig.(8.3) Estimation of Effective Porosity from a Tracer Test.

8.5.2 Estimation of Permeability

Column experiments are performed to estimate the permeability of each type of glass beads. The discharge, Q, through the sample is determined volumetrically by collecting the overflow from the constant head outlet reservoir in a certain period of time. Then the conductivity is calculated by:

$$K = \frac{Q}{A\,J} \tag{8.6}$$

where, K is the hydraulic conductivity, A is the cross section of the column and J is the hydraulic gradient over the sample.

The hydraulic conductivity for various beads are tabulated in Table (8.1) for a dense and loose packing. The hydraulic conductivity values represent a mean of three column experiment measurements per sample and empirical values are estimated using Hazen formula and Kozeny-Carman equation [Bear, 1972]. The empirical values reasonably match the measurements (see Table (8.1)).

8.5.3 Estimation of Dispersivity

For beads size (0.40-0.52 mm) an approximate method is suggested to estimate dispersivity. The method is reliable in the beginning where the plume can be observed clearly. The method is based on measuring roughly the plume outline which is elliptic in shape with maximum length, l_{xx}, and maximum width, l_{yy}, at specified times (see Fig.(8.4)). These dimensions are based on the visual concentration at the plume edges, C_e. C_e is assumed to be equal to 0.01 C_{max} at the plume centre. The concentration field in homogeneous field is known to be Gaussian [Bear, 1972]. According to these assumptions one can write,

$$f\,(l_{xx}/2) = 0.01\,f\,(0) \tag{8.7}$$

where, $f(l_{xx}/2)$ is the ordinate of a normal distribution at location $l_{xx}/2$ which corresponds to C_e and $f(0)$ is the ordinate of the same normal distribution at the centre which corresponds to C_{max}. Eq.(8.7) can be written as,

$$\exp\left[-\frac{l_{xx}^{\,2}}{8\sigma_{xx}^{\,2}}\right] = 0.01 \tag{8.8}$$

and so, one can get the final results for x and y direction as,

$$\begin{aligned}
\sigma_{xx} &\approx \frac{1}{6}\,l_{xx} \\
\sigma_{yy} &\approx \frac{1}{6}\,l_{yy}
\end{aligned} \tag{8.9}$$

Eq.(8.9) is used to estimate the corresponding variance from the plume dimensions.

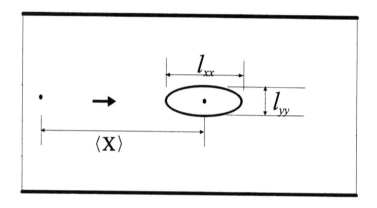

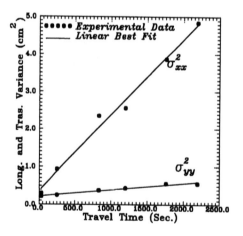

Fig.(8.4) Plume Dimensions and Estimation of Dispersivities.

By plotting the values of σ_{xx}^2 and σ_{yy}^2 versus travel time and using linear best fitting techniques, one can get the slope of the line which corresponds to two times the dispersion coefficient according to [Dagan, 1982]:

$$2D_{xx} = \frac{d\sigma^2_{xx}}{dt} \quad , \quad 2D_{yy} = \frac{d\sigma^2_{yy}}{dt} \tag{8.10}$$

Then, the dispersivities are calculated as,

$$\alpha_l = \frac{D_{xx}}{V}, \quad \alpha_t = \frac{D_{yy}}{V} \tag{8.11}$$

where, α_l is the longitudinal dispersivity, α_t is the transverse dispersivity and V is the mean real velocity given by,

$$V = \frac{K}{\varepsilon} J \tag{8.12}$$

where, J is the gradient on the working section of the model. This method produces dispersivities $\alpha_l=0.333$ mm and $\alpha_t=0.028$ mm. For all glass beads, the formula developed by Harleman, et al. [1963] is used to estimate dispersivities. This formula reads,

$$\frac{D_{xx}}{v_w} = \xi \, R_k^\beta \tag{8.13}$$

where, ξ is an empirical parameter which is 88 for sand grains, and 54 for medium composed of spheres (as in the current case), β is an exponent which is 1.2 for all cases, v_w is the kinematic viscosity of water, and R_k is the Reynolds number of intrinsic permeability defined as,

$$R_k = \frac{V \sqrt{k}}{v_w} \tag{8.14}$$

The intrinsic permeability k is related to the conductivity K by the formula,

$$K = \frac{k \, \rho_w \, g}{\mu_w} \tag{8.15}$$

where, μ_w is the dynamic viscosity of the water, ρ_w density of water and g is the gravitational acceleration equal to 9.81 ms^{-2}.

Form Eq.(8.13), Eq.(8.14) and Eq.(8.15) one obtains the relation between K and α_l, and α_t. The transverse dispersivity, is taken one tenth of the longitudinal one,

$$\alpha_l = \xi \ V^{\beta-1} \ K^{\beta/2} \ v_w^{1-\beta/2} \ g^{-\beta/2}$$
$$\alpha_t = \frac{\alpha_l}{10} \tag{8.16}$$

An estimation of dispersivities by Eq.(8.16) is tabulated in Table (8.1). One may notice that for d_m=0.46 mm the value of α_l obtained from Eq.(8.16) (0.353 mm) is in the same order of magnitude as the value obtained from the visual method described before (0.333 mm). This value is used to validate a homogeneous test under different head gradients and initial concentrations (see section 8.10.1.1).

8.6 Tracer

Potassium Permanganate ($KMnO_4$) solution is used as an optical tracer (dye). It is practically non-reactive. This dye allows us to visually study the evolution of the solute plume. The tracer is injected instantaneously as a pulse through a 1 mm hole half way the height of the box (27.5 cm from the top of the model) at the upstream side (6.5 cm from the inflow side). The dye is introduced via a reservoir connected to the set-up by a plastic tube. The tube is controlled by a valve to provide a certain amount of the dye. The evolution pattern of the plume spreading is monitored photographically at regular intervals.

8.7 Experimental Method To Estimate Concentration Fields

An image analysis approach was applied to provide an accurate detailed concentration field from photographs. A camera with a focal length (35-70 mm) is used. The camera is placed 2 m far from the model. The method of image processing requires control of the room light. For example, to eliminate reflections of the flash from other objects in the room black sheets are placed in the room, on the table where the model is located and behind the model (see Fig.(8.1)).

In the homogeneous test one flash is used. It causes non-uniform light on the model and it is reflected from the top left side as can be seen on the photos in Fig.(8.8). In the heterogeneous test, two flashes are used from both sides of the camera to produce more uniform light intensity on the model and the flashes are placed higher to avoid reflections.

A concentration scale bar is designed to estimate concentrations of the plume.

Four plexiglass boxes (37.3×3.3×1.1 cm³ outer dimensions), consisting of 12 compartments each, are manufactured. Each compartment is 3×3×1 cm³ inner dimensions. The four boxes corresponding to the four glass beads used in the experiments.

A set of concentrations of the dye is prepared. These concentrations ranges from the initial dye concentration, C_o, used at the injection point to zero concentration. These concentrations are poured in each compartment to give the degree of lighting corresponds to each glass with certain concentration. The concentration scale is installed on the model as shown in Fig.(8.5) and Fig.(8.6) to be present in every photograph and to be under the same condition of lighting during the tests. This scale is always prepared before starting each test run. A computation of light intensity-concentration calibration curve can, therefore, be prepared.

Conversion of the intensity of light to a value of solute concentration is an inverse operation. According to the intensity of light at certain location of the plume within the beads one can estimate the corresponding concentration from the concentration scale bar.

8.8 The Experimental Program

In each test run, the set-up is packed and sealed. Then flow is introduced by imposing a certain gradient on the model. After steady state condition is established (the piezometers stopped from raising or falling) a known volume of dye is injected into the model to reproduce a pollution source. After the dye starts to spread in the flow, photographs of the evolution of the solute plume are taken. Analysis of the experimental measurements (piezometric heads at some locations in the flow field, the volume of the injected dye, plume centroid, plume dimensions, and time of each snapshot) and photographs are taken. The experiments involved homogeneous, and heterogeneous media. For each of the two media, between 2 experimental trials were run with initial concentrations, C_o, ranging from 0.00316 gram/cm³ to 0.0104 gram/cm³ $KMnO_4$. The injected mass is related to the initial concentration by

$$M_o = V_o\, C_o \qquad (8.17)$$

where, M_o is the injected mass of solute, V_o is the injected volume of dye and C_o is the initial dye concentration.

For the experiments in the homogeneous medium, the box is filled with size (d_m=0.46 mm) glass beads. For the experiments in the heterogeneous media, the structures are displayed in Fig.(8.5) and Fig.(8.6) for medium 1 and medium 2 respectively.

8.9 Numerical Simulation

Numerical simulations have been carried out to investigate the transport behaviour. The simulations are preformed by various computer models developed in this study. The simulation steps followed are:

(i) Schematization:
The heterogeneous medium 1 and medium 2 respectively are schematized and displayed in Fig.(8.5) and Fig.(8.6).

(ii) Discretization and Parameter Assignation:
The homogeneous medium is discretized with gridsize 1×1 cm. For heterogeneous media a mesh of 2.5×1 cm is used. The parameters of each class of glass beads are assigned according to Table (8.1). An average permeability of the loose and dense values are used. Local dispersivities are estimated using Eq.(8.16).

(iii) Boundary and Initial Conditions
The boundary conditions are constant head boundary at left and right sides of the flow domain. These heads are measured by piezometers number 2 and number 5 for left and right boundary respectively (see Fig.(8.1)). No-flow boundaries are assigned to the top and bottom of the model. For the initial condition of the transport problem, an instantaneous injection is proposed from a single point in Fig.(8.5) or from multiple-points as in Fig.(8.6).

(iv) Flow Simulator
The flow problem of each test is solved in terms of potentials and stream functions. The results are displayed in Fig.(8.5) and Fig.(8.6). The potentials are checked with measured locations which are presented in Table (8.2), Table (8.3) and Table (8.4).

(v) Transport Simulator
The transport model developed in Chapter 6 is used for solving the transport problem.

8.10 Simulation Results and Discussion

8.10.1 Deterministic Simulations

Some simulations have been performed to validate the numerical models. If the heterogenous permeability arrangements are known with certainty a deterministic simulation is followed. This simulation can be viewed as single realization approach (Chapter 6).

The numerical model reproduces the flow pattern in a satisfactory way. The heads at some fixed locations along the flow are measured by some piezometers.

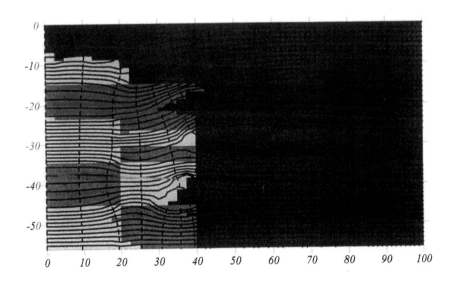

Fig.(8.5) Schematization of Heterogeneous Medium (1) with Flownet.

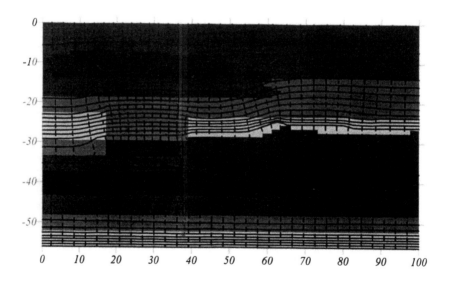

Fig.(8.6) Schematization of Heterogeneous Medium (2) with Flownet.

8.10.1.1 Homogeneous Medium

For the homogeneous medium, the results are displayed in Fig.(8.7) for plume spatial moments. Fig.(8.8), Fig.(8.9) and Fig.(8.10) show snapshots of the plume at different times. The expected spread of the plume is clearly seen. The highest concentrations occur in the central part of the plume, while the outer edges are more diluted. The numerical simulation matches the experimental observations reasonably well (flow and transport). Table (8.2) shows the measured heads at some locations and the corresponding values obtained from the numerical simulations.

Table (8.2) Measured and Simulated Heads in Homogeneous Medium.

Piezometer Fig.(8.1)	Location (cm)	Measured Head (cm)	Simulated Head (cm)
no. 3	X = 36.8	44.10	44.14
no. 4	X = 63.5	40.50	40.40

The movement of the centroid of the plume and the longitudinal and transverse variances are presented in Fig.(8.7). Numerical simulation are performed with the dispersivities calculated by the suggested visual method.

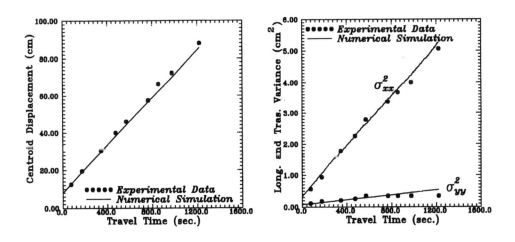

Fig.(8.7) Comparison of Plume Statistics in Homogeneous Medium.

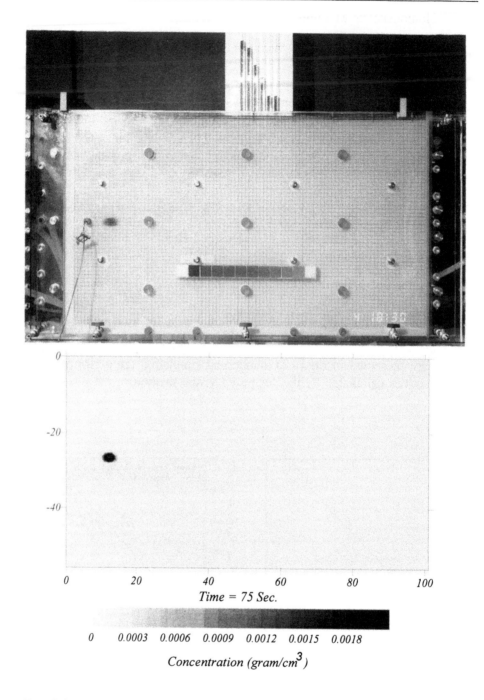

Time = 75 Sec.

Concentration (gram/cm³)

Fig.(8.8) Snapshot (1) of The Plume in Homogeneous Medium (Experimental and Numerical Results).

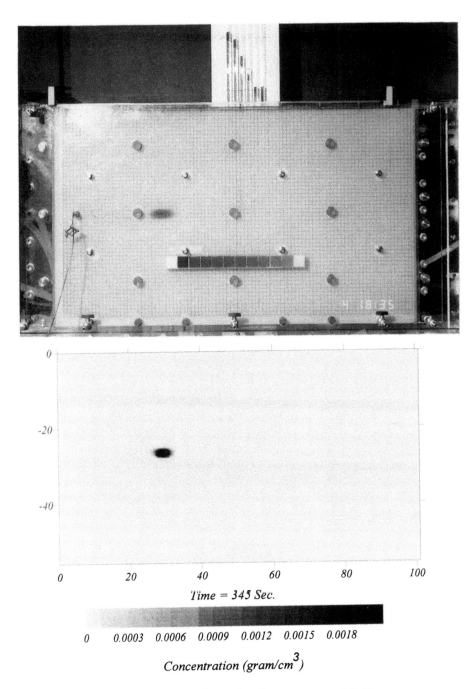

*Fig.(8.9) Shapshot (2) of The Plume in Homogeneous Medium
(Experimental and Numerical Results).*

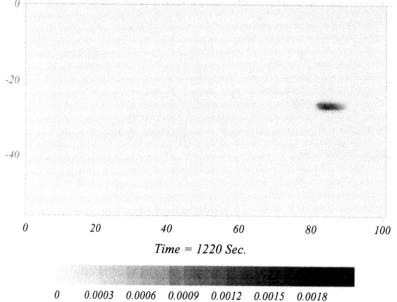

Time = 1220 Sec.

0   0.0003   0.0006   0.0009   0.0012   0.0015   0.0018

Concentration (gram/cm^3)

Fig.(8.10) Snapshot (3) of The Plume in Homogeneous Medium (Experimental and Numerical Results).

8.10.1.2 Heterogeneous Media

In these experimental tests the simulation is restricted to heterogeneous structures which can be constructed in the experimental set-up. Fig.(8.5) and Fig.(8.6) show the heterogeneous structures in Medium 1 and 2 respectively. In the experiments each zone is composed of a glass beads having a distinct conductivity and particle diameter. Due to the distinction in conductivity there is a corresponding distinction in local dispersivities. The values of conductivities used in the numerical simulations are given in Table (8.1) and the local dispersivities are estimated by Eq.(8.16).

(i) Heterogeneous Medium 1 with Single Point Instantaneous Injection
The flow results show good agreements in terms of the head at some piezometers located in the flow field. Table (8.3) shows the measured and simulated heads at two piezometer locations. The simulated flow field superimposed over the heterogeneous structure is presented in Fig.(8.5).

For the transport simulations, Fig.(8.11), Fig.(8.12) and Fig.(8.13) show three snapshots of the solute plume at certain time intervals. The overall comparison between the laboratory tests and the numerical simulation is satisfactory. Again, the highest concentrations occur in the middle part, while the outer edges are more diluted.

Fig.(8.14) shows the plume statistics. It is clear from the numerical simulations that the centroid displacement is linear, and the breakthrough curves display Gaussian characteristics. However, the spreading of the plume shows non-Fickian behaviour. Snapshot (3) in Fig.(8.13) shows more longitudinal spreading than the simulation. This is due to the fingering which shows apparently more spreading in the longitudinal direction and is not represented by the advection dispersion equation defined over represented elementary volume, *REV*. A subdiffusive regime can be noticed at the end of the curve.

One may recognize that the longitudinal dispersion is decreasing. This behaviour is due to the fact that the plume size is small with respect to the heterogeneous structure. The plume movement is influenced by the spatial arrangements of the various classes of glass beads while the spreading mechanisms are governed by

Table (8.3) Measured and Simulated Heads in Heterogeneous Medium (1).

Piezometer Fig.(8.1)	Location (cm)	Measured Head (cm)	Simulated Head (cm)
no. 3	X = 36.8	47.00	46.75
no. 4	X = 63.5	40.80	40.90

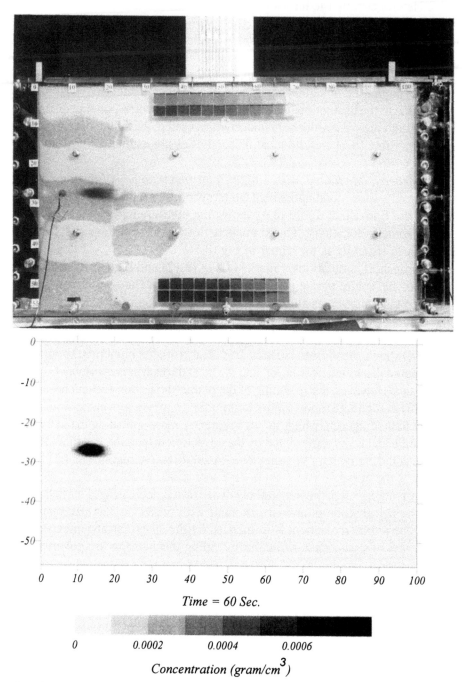

Fig.(8.11) Snapshot (1) of The Plume in Heterogeneous Medium 1 (Experimental and Numerical Results).

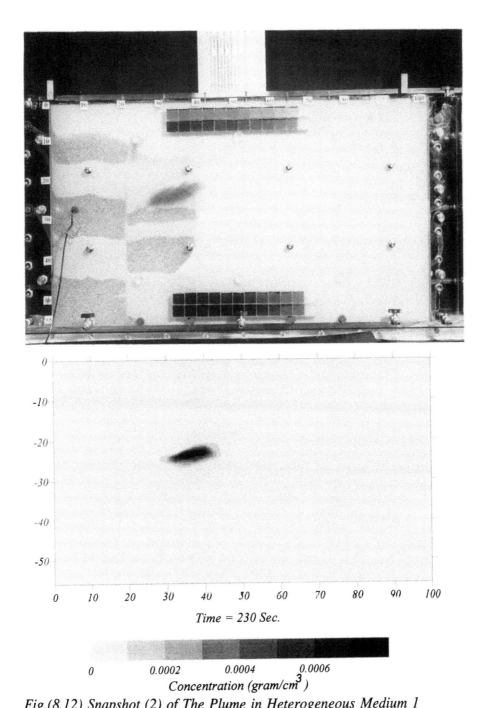

Fig.(8.12) Snapshot (2) of The Plume in Heterogeneous Medium 1 (Experimental and Numerical Results).

267

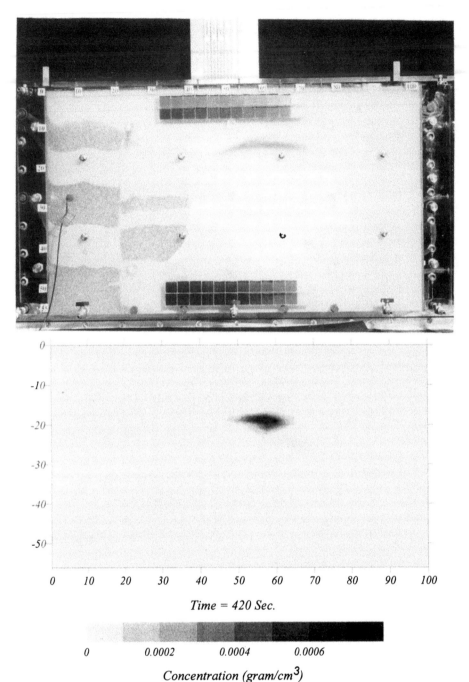

Time = 420 Sec.

Concentration (gram/cm³)

Fig.(8.13) Snapshot (3) of The Plume in Heterogeneous Medium 1 (Experimental and Numerical Results).

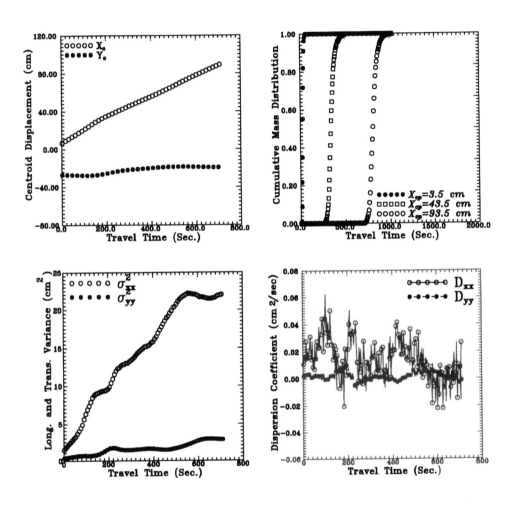

Fig.(8.14) Plume Statistics in Heterogeneous Medium (1).

the local dispersivities in each class of glass beads. Since the plume moves from a highly permeable zone near the source to a less permeable zones the spreading is high in the beginning. After some certain time the plume goes to zones of a lower dispersion. As a consequence, the rate of spreading decreases which causes the longitudinal variance to grow slowly and subdiffusive regime occurs.

(ii) Heterogeneous Medium 2 with Multi-Points Instantaneous Injection
In this test, heterogenous structure is used in the form of perfect and imperfect horizontal layers (see Fig.(8.6)). The simulated flow field is displayed in the same figure. The measured and simulated heads at some locations are shown in Table (8.4). The agreement is satisfactory.

Table (8.4) Measured and Simulated Heads in Heterogeneous Medium (2).

Piezometer Fig.(8.1)	Location (cm)	Measured Head (cm)	Simulated Head (cm)
no. 3	X = 36.8	40.40	40.80
no. 4	X = 63.5	38.20	38.60

Fig.(8.15), Fig.(8.16) and Fig.(8.17) show snapshots of the plume at different times. The results show general agreement in terms of plume shapes at the specified times. Fig.(8.18) displays the plume centroid, breakthrough curves, longitudinal and lateral spreading variances and dispersion coefficients.

The breakthrough curves show non-Gaussianity due to the influence of low permeable zones where the part of the plume is trapped and takes a longer time to be released into the high permeable zones. The longitudinal variance shows a superdiffusive regime at early times. However, at 250 seconds it seems that it is reaching a dispersive regime. This behaviour can be attributed to the fact that the plume sampled all the heterogeneity in the system so the heterogeneity does not contribute any further in its spreading. Therefore Fickian regime is reached with ($D_{xx} = 2.5$ cm^2/sec).

In heterogeneous medium 1 the breakthrough curve shows Gaussianity. However, in heterogeneous medium 2 the breakthrough curve shows non-Gaussianity. That mean the breakthrough curve does not always reveal the real dispersive behaviour. It is not a reliable measure of transport characteristics in heterogeneous media.

8.10.2 Stochastic Simulations

So far, deterministic simulations have been carried out to study the dispersion mechanisms which take place in heterogeneous medium. In this section some stochastic simulations are carried out. The idea behind these simulation is to estimate

Fig.(8.15) Snapshot (1) of The Plume in Heterogeneous Medium 2 (Experimental and Numerical Results).

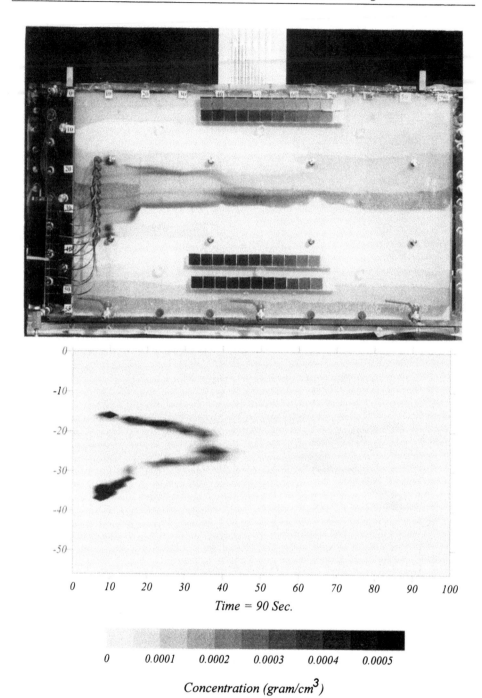

Time = 90 Sec.

Concentration (gram/cm³)

Fig.(8.16) Snapshot (2) of The Plume in Heterogeneous Medium 2 (Experimental and Numerical Results).

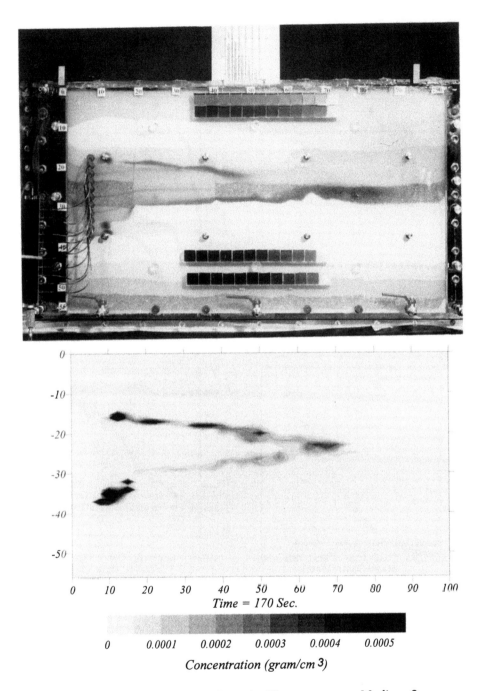

Time = 170 Sec.

Concentration (gram/cm³)

Fig.(8.17) Snapshot (3) of The Plume in Heterogeneous Medium 2
(Experimental and Numerical Results).

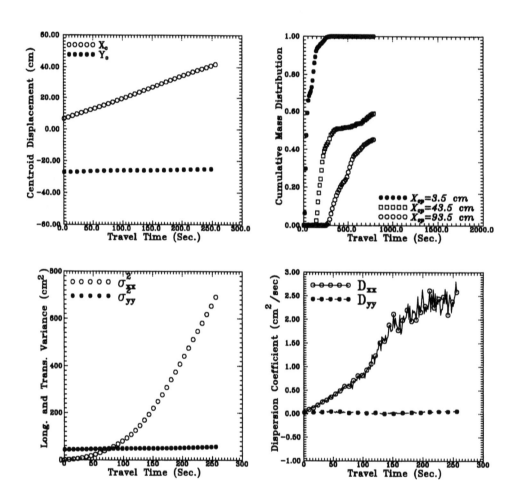

Fig.(8.18) Plume Statistics in Heterogeneous Medium 2.

the uncertainty in the plume spreading due to the uncertainty in the heterogeneous structure which is often the case in field problems where heterogeneity is known only at sampled locations.

The purpose of these simulations, as described in detail in Chapter 7, is to deal with prediction uncertainty to quantify the uncertainty which is associated with predicting the concentration fields and/or plume spatial moments.

(i) Heterogeneous Medium 1 with Single Point Instantaneous Injection
The simulations assume that the heterogeneous arrangements of the permeability is known in terms of transition probabilities between the classes of glass beads (states) that constitute the system. Fig.(8.19) shows medium 1 and its stochastic simulation. Flow and transport have been solved in 100 realizations of these artificial structures with similar statistics. The ensemble concentration fields and the standard deviation in concentrations are calculated and displayed in Fig.(8.20), Fig.(8.21) and Fig.(8.22) at the same snapshots given in Fig.(8.11), Fig.(8.12) and Fig.(8.13) respectively. The ensemble concentrations show a wide spread over the region. This is due to the fact that there is a large degree of freedom in the variability from one realization to another. The considerable uncertainty in the plume locations reflects this freedom.

When the spatial moments of the real plume are compared with the ensemble mean and 95% confidence intervals, Fig.(8.23), the spatial moments fall in the 95% confidence intervals because the envelope of uncertainty is wide and the simulations take place only within 125 sec.

(ii) Heterogeneous Medium 2 with Multi-Points Instantaneous Injection
The same procedure has been followed for the heterogeneous test. Fig.(8.24) shows medium 2 and its stochastic simulation. Flow and transport have been solved in 100 realizations of these artificial structures with similar statistics.

The ensemble concentration fields and the standard deviation in concentrations are calculated and displayed in Fig.(8.25), Fig.(8.26) and Fig.(8.27) at the same snapshots given in Fig.(8.15), Fig.(8.16) and Fig.(8.17) respectively. The ensemble concentrations show a wide spread over the region. The regions of high uncertainty are located near the source. This is due to the fact that there is a large degree of freedom in the variability from one realization to another.

Again, comparing the plume spatial moments of the real plume with the ensemble mean and 95% confidence intervals, Fig.(8.28), shows that the single realization falls in the 95% confidence interval because the envelope of uncertainty is wide. However, for the macro-dispersion coefficients the single realization falls on the upper boundary of the interval. This is due to the fact that the 95% confidence interval of the spatial moments is independent beyond the 95% confidence intervals of the macro-dispersion coefficients. The envelop in the longitudinal macro-dispersion is dominated by local variations (noise) in the longitudinal variance.

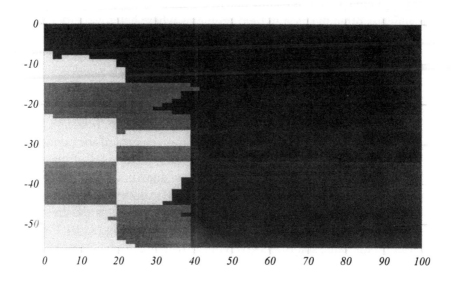

(a) Schematization of Heterogeneous Medium 1.

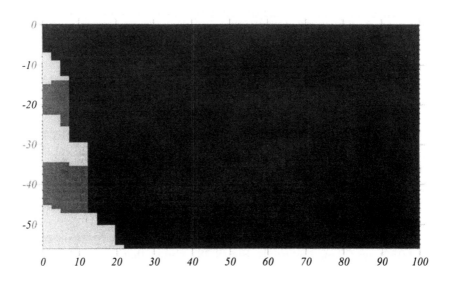

(b) Stochastic Simulation.

Fig.(8.19) Heterogeneous Medium 1 and Its Stochastic Simulation.

Ensemble Concentration (gram/cm^3)

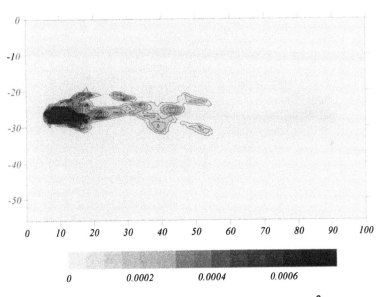

Standard Deviation in Concentration (gram/cm^3)

Time = 60 Sec.

Fig.(8.20) Snapshot (1) of The Ensemble Plume and Standard Deviation in Heterogeneous Medium 1.

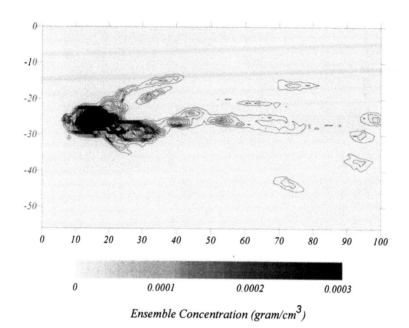

Ensemble Concentration (gram/cm³)

Standard Deviation in Concentration (gram/cm³)

Time = 230 Sec.

Fig.(8.21) Snapshot (2) of The Ensemble Plume and Standard Deviation in Heterogeneous Medium 1.

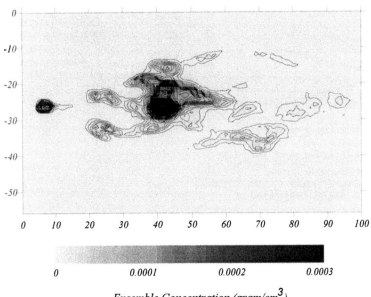

Ensemble Concentration (gram/cm^3)

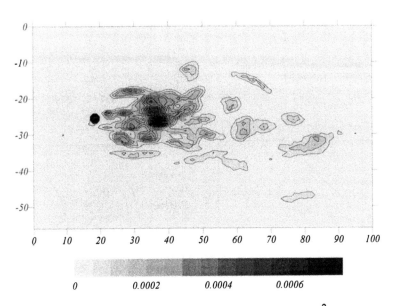

Standard Deviation in Concentration (gram/cm^3)

Time = 420 Sec.

Fig.(8.22) Snapshot (3) of The Ensemble Plume and Standard Deviation in Heterogeneous Medium 1.

279

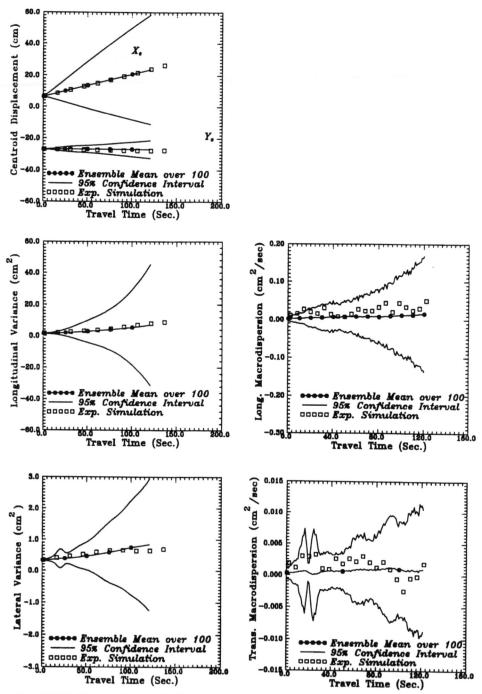

Fig.(8.23) Comparision of The Deterministic Simulation with The Stochastic Simulation in Terms of Plume Statistics (Medium 1).

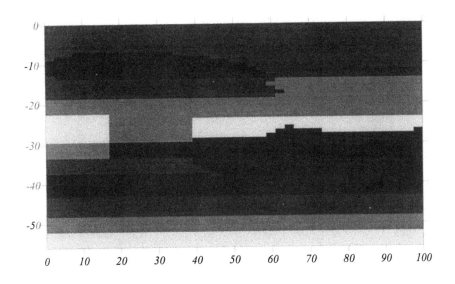

(a) Schematization of Heterogeneous Medium 2.

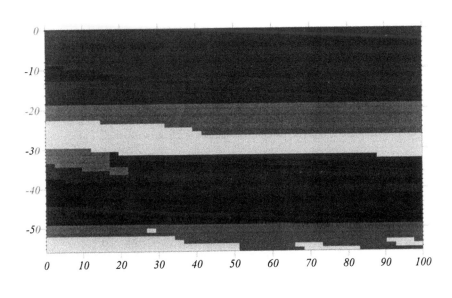

(b) Stochastic Simulation.

Fig.(8.24) Heterogeneous Medium 2 and Its Stochastic Simulation.

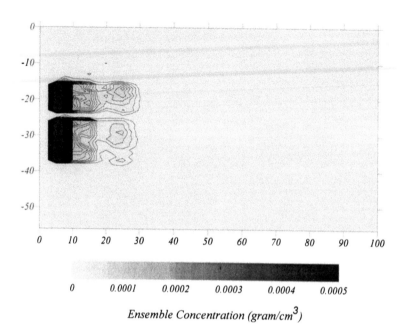

Ensemble Concentration (gram/cm^3)

Standard Deviation in Concentration (gram/cm^3)

Time = 50 Sec.

Fig.(8.25) Snapshot (1) of The Ensemble Plume and Standard Deviation in Heterogeneous Medium 2.

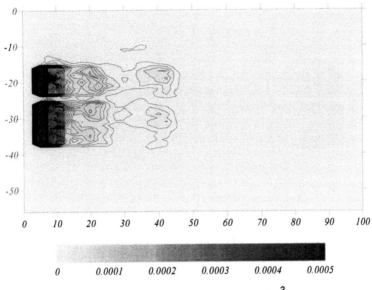

Ensemble Concentration (gram/cm³)

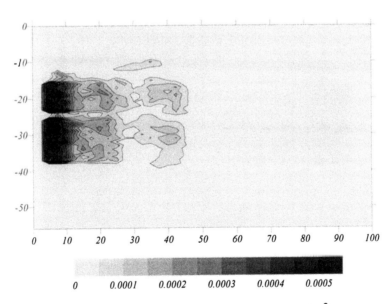

Standard Deviation in Concentration (gram/cm³)

Time = 90 Sec.

Fig.(8.26) Snapshot (2) of The Ensemble Plume and Standard Deviation in Heterogeneous Medium 2.

283

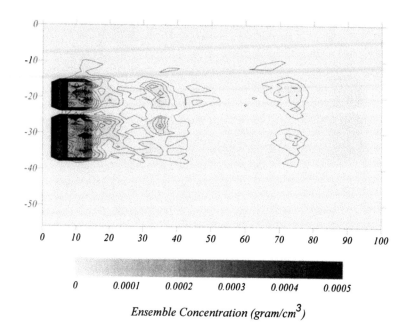

Ensemble Concentration (gram/cm^3)

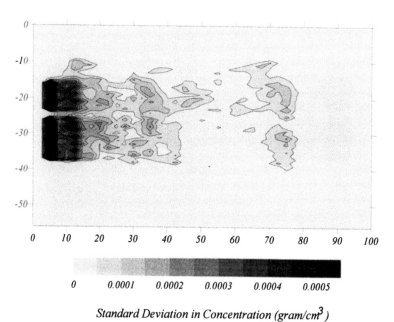

Standard Deviation in Concentration (gram/cm^3)

Time = 170 Sec.

Fig.(8.27) Snapshot (3) of The Ensemble Plume and Standard Deviation in Heterogeneous Medium 2.

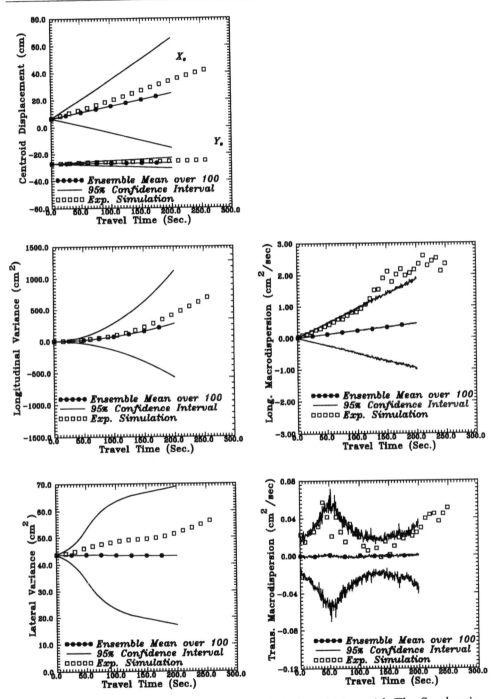

Fig.(8.28) Comparision of The Deterministic Simulation with The Stochastic Simulation in Terms of Plume Statistics (Medium 2).

8.11 Summary and Conclusions

A laboratory study on solute transport in artificially homogeneous and heterogeneous media is carried out. The experiments are performed in a two-dimensional box (Fig.(8.1)). The porous media are constructed from glass beads with different sizes (see Table (8.1) and Fig.(8.2)). The tracer utilized is a Potassium Permanganate ($KMnO_4$) solution. The investigated media are homogeneous medium and two heterogeneous media (medium 1 and 2). The experiments demonstrate the influence of the presence of preferred pathways, medium contrasts, the initial plume size and the spatial arrangements of the less and highly permeable zones on dispersion characteristics.

Because of the time difficulties faced to use image processing to fully characterize concentration fields, (it was actually out of the scope of this study) a qualitative analysis has been followed. From qualitative analyses of the experimental data and some numerical deterministic and stochastic simulations of the tests the following conclusions and remarks can be made:

(1) Although the geological interfaces, which causes mega-dispersion, generally cannot be simulated easily in laboratory models, the tests illustrate the usefulness of controlled experiments in supporting theories of solute transport in heterogenous media.

(2) The tests demonstrate that it is possible to get better insight into the transport behaviour in homogeneous and heterogeneous media. The influence of various interfaces on dispersion behaviour can be recognized. Subdiffusive dispersion is observed in medium 1 because the plume moves from highly permeable to less permeable zones. Superdiffusive dispersion can be seen in medium 2 due to the existence of preferential flow paths.

(3) The deterministic numerical simulations of the physical tests show qualitatively similar results in terms of plume spreading.

(4) The scale-dependent dispersion coefficient can be seen from the simulations in heterogeneous medium 2 (see Fig.(8.18)).

(5) When the plume is small in comparison with the size of the heterogeneity (heterogeneous medium 1) the plume movement is influenced by the spatial arrangements of the various units but the spreading mechanisms are mainly governed by the local dispersion process inside each unit. In other words, the heterogeneity does not contribute in the spreading mechanisms it only influences the path of the plume.

(6) Both Gaussian (symmetric) and non-Gaussian breakthrough curves (steep fronts at the early times and elongated tails at later times) can be found in heterogeneous media. The condition of occurrence of Gaussian or non-Gaussian behaviour depends on the extension of the source perpendicular to the flow direction and the spatial arrangements of the heterogeneity.

(7) Good predictions of solute plumes are only possible by performing deterministic simulations using random walk method if all details of the aquifer heterogeneity and initial conditions are known in a deterministic sense.

(8) Stochastic simulations related to uncertainty in system descriptions which is characterized in terms of transition probabilities show considerable uncertainty of the model predictions due the large degree of freedom from one realization to another. The only way to reduce the uncertainty in field applications is to perform conditioning on soft and hard data as explained in Chapter 7. For example by giving a 'weight' to each realization, where the weight is defined by a geologist or a geohydrologist (engineering intuition, knowledge of deposition history, experience, etc.).

(9) The stochastic numerical simulations of the physical experiments emphasized that the ensemble transverse macro-dispersion coefficient is practically zero and has therefore no relation with the transverse micro-scale dispersion coefficient. This result underscores the growing belief that the effective macro-dispersion has no transverse component.

General Conclusions and Recommendations

9.1 Conclusions

Grounwater flow and contaminant transport in aquifers are strongly influenced by the geological heterogeneity at various scales. This research has led to a better conceptual understanding of the transport mechanisms in various types of heterogeneous formations: single scale heterogeneity such as stationary continuous Gaussian and discrete Markovian fields and multiple scale heterogeneity such as compound non-stationary fields (Gaussian-Markovian fields). The impacts of the newly developed tools of heterogeneous characterization are illustrated by extensive single and multiple-realization numerical simulations and laboratory experiments on flow and non-reactive solute transport. The developed coupled Markov chain methodology provides a suitable tool in quantifying uncertainty of the geometrical configuration of the formation. The work highlights also the incorporation of both geological (geometrical configuration) and parametric uncertainties on transport predictions.

The following main conclusions can be drawn from this study:

(1) The present research demonstrates that the coupled Markov chain model (Chapter 3) is capable of characterizing various geological configurations: perfectly and imperfectly stratified patterns with different thicknesses, inclusions, horizontal and inclined bedding with different angles of inclinations and large scale geological units.

(2) The suggested methodology can address parametric variability. The spatial structure is characterized by conditional probabilities which are more interpretative in a geological sense and easier to estimate from soft geological information and the available hard data than auto-covariance or variogram functions. The methodology is more general than the one used by Krumbein [1967] since it simulates a two or three-dimensional field.

288

(3) The proposed hybrid model (Chapter 5) which is composed of the coupled Markov model and the turning bands algorithm can be used to generate correlated spatial distributions of conductivity within the geological structure, with all realizations satisfying the soft geological knowledge (reproduction of the transition probabilities) and hard data (reproduction of the *PDFs* and its moments). The representation of the spatial variability by this model is promising and it provides a good step for introducing multiple scale heterogeneity (at least for the macroscopic and the megascopic scale).

(4) From some synthetic examples, it has been shown that high contrast in the mean parameter value of various individual units is the main factor of pronounced non-stationarity in comparison with other types of non-stationarity. The assumption of a global stationary random field is not adequate to describe many heterogeneous fields as far as the non-stationarity in the mean is concerned.

(5) In the case of relatively low conductivity contrast of the individual layers, it has been shown that there is no difference between fitting normal or lognormal density functions to the probability density functions, *PDF*, of the effective conductivity from the numerical Monte-Carlo flow simulations (Chapter 4) which are performed to estimate the effective conductivity of heterogeneous media characterized by the coupled Markov model. However, in case of relatively high contrast the lognormal density fits better. This is in accordance with the general belief of the lognormality of the *PDF* of the conductivity. The hydraulic anisotropy of the effective conductivity is clearly reflected by the anisotropy in the heterogeneous structure.

(6) From the single-realization numerical simulations one may conclude that the effects of the internal variability within the geological units are more pronounced on the transport characteristics than on the hydrodynamics of flow.

(7) Various transport mechanisms (Fickian: diffusive dispersion, non-Fickian: sub-diffusive dispersion and superdiffusive dispersion) are observed in the simulations. These mechanisms are controlled by the spatial arrangements of the various geological units, by the extension of the source perpendicular to the global flow direction, and by the method of characterization of the heterogeneous medium. This observation explains the limitations of the conventional simulation methods.

(8) The concentration distribution in many geological setting is not Gaussian. This particularly happens when high heterogeneity contrast and preferential flow paths are present in the system. For field applications, the choice of the appropriate method of characterization, such as Gaussian, Markovian, compound or otherwise, should be tested carefully.

(9) From an experiment on a perfectly stratified formation in the absence of pore scale dispersivities it has been shown that the internal variability (macro heterogeneity) within the layers mimics the influence of pore scale dispersion process. That means, for a practical point of view, it is difficult to interpret results of a field tracer test.

(10) The experiments showed intermittent chains of Fickian and non-Fickian regimes in many geological deposits. That means, distinct effective macro-dispersivities do not exist.

(11) Irrespective the different regimes of flow and transport mechanisms in the heterogeneous media it has been shown (Chapter 4) that a characteristic macro-effective conductivity exists.

(12) From the deterministic numerical simulations of the laboratory experiments (Chapter 8) good predictions of the evolution of solute plumes are found using the random walk method, provided that all details of the aquifer heterogeneity, and the initial and boundary conditions are known with certainty.

(13) The experiments with a small-size plume in comparison with the size of the heterogeneity (medium 1) yield: the plume movement is influenced by the spatial arrangements of the various units, while the spreading mechanisms are mainly governed by the local micro-dispersion process inside each unit. That means, the heterogeneity does not contribute in the mixing process; it only influences the centroid movements.

(14) The simulated breakthrough curves of the laboratory experiments show in some experiments Gaussian characteristics and in some other experiments non-Gaussian characteristics. This means that the breakthrough curve is not a reliable measure of transport characteristics in heterogeneous deposits.

(15) The multi-realization numerical experiments, Monte-Carlo approach, (Chapter 7) and the stochastic numerical simulations of the laboratory experiments (Chapter 8) demonstrate the potential applicability of the coupled Markov chain model in quantifying geological uncertainty when the system possesses a Markovian property. Many geological deposits do possess such a property [Krumbein, 1967].

(16) The results indicate that a relatively high uncertainty will often be associated with the ensemble mean as a predictor of the concentration distribution. The greatest uncertainty, which represents the worst case in predictions, occurs when information is lacking about the geological setting and parametric values of each geological unit. This uncertainty is due to the large degree of freedom from one realization

290

to the other. It has been found that conditioning on the geological configuration reduces this uncertainty significantly. Therefore, a prior and classical geological knowledge is most important. It can be incorporated in the methodology to weight the various realizations and to select the most probable ones.

(17) In case of relatively low contrast in conductivity one may conclude that there is no remarkable influence on plume spatial moments and the evolution of their uncertainty under geological and both geological and parametric uncertainty. However, in the case of relatively high contrast in conductivity it has been demonstrated that there is a clear retardation of the plume spatial moments and a retardation in the evolution of their uncertainty for both geological and parametric uncertainty. This is due to the influence of the lognormality of the *PDF* of conductivity.

(18) An actual geological structure (single realization) gives a significantly different dispersion behaviour from the ensemble mean over many, say 100, realizations. It has been found that the contaminated area can be indicated with certainty (described by the imaginary envelope of all possible realizations), however the value of concentration at certain location is subjected to a high degree of uncertainty. This result is of great practical value.

(19) The stochastic numerical experiments of the laboratory tests (Chapter 8) emphasized that the ensemble transverse macro-dispersion coefficient is practically zero and has therefore no relation with the transverse micro-scale dispersion coefficient. This result underscore the growing belief that the effective macro-dispersion has no transverse components.

(20) The coupled Markov model can in principle be incorporated quite easily in existing groundwater quality management models to provide insight in the effect of geological uncertainty.

9.2 Recommendations

(1) Natural extension of the coupled Markov theory to three-dimensions is a simple extension; just one extra transition probability matrix will be used to account for the transitions in the third direction.

(2) Conditional simulation using information from various bore loggings can be taken into account to reduce the uncertainty of the generated image.

(3) The coupled Markov chain model presented in this thesis assumes that the transition probabilities are stationary (do not change with translations). A departure from this assumption would be useful to characterize more complex geological deposits where the transition probabilities change with location.

(4) The laboratory experiments have underscore the results found in this study. However, there is a need for large scale field characterization to validate the hybrid model at the macroscopic scale (the scale of variability of hydrogeological parameters) and the megascopic scale (the scale of variability of the lithological units).

(5) It is recommended to extend the present transport model for chemical processes, sorption, desorption, retardation, decay and biological transformations. This extension seems straightforward, without serious complications.

(6) In this thesis it has been considered that the hydraulic conductivity and the geological structure are the only source of uncertainty. For field applications it is of more interest to consider the uncertainty of other hydrogeological parameters and of boundary and initial conditions, as well. The presented new method can easily be adapted for this purpose.

References

Ababou, R. & Gelhar, L.W. (1990). Self-similar Randomness and Spectral Conditioning: Analysis of Scale Effect in Subsurface Hydrology. In *Dynamics of Fluids in Hierarchical Porous Formations*. Edited by J. Cushman, pp 393-428, Academic Press. London.

Ababou, R. & McLaughlin, D. & Gelhar, L.W. & Tompson, A.T.B. (1989). Numerical Simulation of Three-dimensional Saturated Flow in Randomly Heterogeneous Porous Media. *Transport in Porous Media*. 4: 549-565.

Agterberg, F.P. (1974). *Geomathematics: Developments in Geomathematics 1*. Elsevier Scientific Publishing Co., Amsterdam, The Netherlands.

Akker, C. van den (1982). *Numerical Analysis of The Stream Function in Plane Groundwater Flow*. Ph.D. Thesis. Delft University of Technology. The Netherlands.

Baker, R. (1984). Modelling Soil Variability as a Random Field. *Mathematical Geology*. 16(5): 435-448.

Bakr, A.A. (1976). *Stochastic Analysis of The Effects of Spatial Variations of Hydraulic Conductivity on Groundwater Flow*, Ph.D. dissertation, New Mexico Institute of Mining and Technology, Socorro.

Bakr, A.A., Gelhar, L.W., Gutjahr, A.L. & MacMillan, J.R. (1978). Stochastic Analyses of Spatial Variability in Subsurface Flows. 1. Comparison of One and Three-dimensional Flows. *Water Resour. Res.* 14(2):263-271.

Bartlett, M.S. (1975). *The Statistical Analysis of Spatial Pattern*. Chapman and Hall, London.

Bear, J. (1961a). Some Experiments in Dispersion. *J. Geophys. Res.*, 66(8): 2455-2467.

Bear, J. (1961b). On The Tensor Form of Dispersion. *J. Geophys. Res.*, 66(4): 1185-1197.

Bear, J. (1972). *Dynamics of Fluids in Porous Media*. American Elsevier, New York.

Bear, J. (1979). *Hydraulics of Groundwater*, McGraw-Hill. New York.

Bear, J. & Verruijt, A. (1987). *Modelling Groundwater Flow and Pollution, with Computer Programs for Sample Cases*. Reidel, Dordrecht.

Bennett, R.J. (1979). *Spatial Time Series*. Pion Limited, 207 Brondesbury Park, London.

Beran, M.J. (1968). *Statistical Continuum Theories*. Interscience, New York.

Billingsley, P. (1995). *Probability and Measure*. Third Edition. A Wiley-Intersience Publication. John Wiley & Sons.

Brannan, J.R. & Haselow, J.S. (1993). Compound Random Field Models of Multiple Scale Hydraulic Conductivity. *Water Resour. Res.* 29(2), pp. 365-372.

Bras, R.L. & Rodriguez-Iturbe, I. (1984) *Random Functions and Hydrology*. Addison-Wesley publishing company, Inc.

Brook, D. (1964). On The Distinction Between The Conditional and Joint Probability Approaches in The Specification of Nearest Neighbour Systems. *Biometrika.* 51, 481-483.

Boggs, J.M. & Young, S.C. & Beard, L.M. & Gelhar, L.W. & Rehfeldt, K.R. & Adams, E.E. (1992). Field Study of Dispersion in a Heterogeneous Aquifer, 1, Overview and Site Description. *Water Resour. Res.,* 28, pp 3281.

Borgman, L. & Taheri, M. & Hagan, R. (1984). Three-dimensional Frequency-domain Simulation of Geological Variables. In Verly, G. David, M. Journel, A.G. and Marechal, A. (Eds.), *Geostatistics for Natural Resource Characterization,* Part 2: Reidel, Dordrecht, Netherlands.

Brower, H. (1978). *Groundwater Hydrology.* New York, McGraw-Hill, 480p.

Burrough, P.A. (1983a). Multiscale Sources of Spatial Variation in Soil; I. The Application of Fractal Concepts to Nested Levels of Soil Variation. *J. of Soil Science* 34, pp. 577-597.

Burrough, P.A. (1983b). Multiscale Sources of Spatial Variation in Soil; II. A Non-Brownian Fractal Model and Its Application in Soil Survey. *J. of Soil Science* 34, pp. 599-620.

Chessa, A. (1995). *Conditional Simulation of Spatial Stochastic Models for Reservoir Heterogeneity.* Ph.D. Thesis. Delft University of Technology. The Netherlands.

Christakos, G. (1992). *Random Fields Models in Earth Sciences.* Academic, San Diego, Calif.

Christakos, G. & Hristopulos, D.T. & Miller, C.T. (1993). Stochastic Diagrammatic Analysis of Groundwater Flow in Heterogeneous Porous Media. *Water Resour. Res.* 31(7), pp. 1687-1703.

Chirlin, G.R. & Dagan, G. (1980). Theoretical Head Variograms for Steady Flow in Statistically Homogeneous Aquifers. *Water Resour. Res.,* 16(6), pp. 1001-1015.

Clifton, P.M. & Neuman, S.P. (1972). Effect of Kriging and Inverse Modelling onConditional Simulation of Avra Valley Aquifer in Southern Arizona. *Water Resour. Res.* 18(4): 1215-1234.

Cross, G.R. & Jain, A.K. (1983). Markov Random Field Texture Models. *IEEE Transactions on Pattern Analysis and Machine Intelligence.* Vol. PAMI-5, no.1.

Cushman, J.H. & Ginn, T.R. (1993). Nonlocal Dispersion in Media with Continuously Evolving Scales of Heterogeneity. *Trans. Porous Media.* Vol. 13. pp. 123-138.

Dagan, G. (1979). Models of Groundwater Flow in Statistically Homogoeneous Porous Formations. *Water Resour. Res.* 15(1): 47-63.

Dagan, G. (1981). Analysis of Flow Through Heterogeneous Random Aquifers by The Method of Embedding Matrix, 1. Steady flow. *Water Resour. Res.* 17(1): 107-121.

Dagan, G. (1982a). Analysis of Flow Through Heterogeneous Random Aquifers, 2. Unsteady Flow in Confined Formations. *Water Resour. Res.* 18(5): 1571-1585.

Dagan, G. (1982b). Stochastic Modelling of Groundwater Flow by Unconditional and Conditional Probabilities, 2. The Solute Transport. *Water Resour. Res.*18(4):835-848.

Dagan, G. (1984). Solute Transport in Heterogeneous Porous Formations. 45: 151-177.

Dagan, G. (1985). Stochastic Modeling of Groundwater Flow by Unconditional and Conditional Probabilities: The Inverse Problem *Water Resour. Res.* 21(1): 65-72.

Dagan, G. (1986). Statistical Theory of Groundwater Flow and Transport: Pore to Laboratory. Laboratory to Formation and Formation to Regional Scale. *Water Resour. Res.* 22(9): 120S-134S.

Dagan, G. (1989). *Flow and Transport in Porous Formations,* Springer, Verlag, Berlin.

Dagan, G. (1994). The Significance of Heterogeneity of Evolving Scales to Transport in Porous Formations, *Water Resour. Res.,* 30(12), pp 3322-3336.

Davis, J.C. (1973). *Statistics and Data Analysis in Geology.* J. Wiley, New York.

Davis, M.W. (1987). Production of Conductional Simulations Via The LU Decomposition of The Covariance Matrix. *Math. Geol.* Vol. 19, no. 2, pp 91-98.

Delhomme, J.P. (1979). Spatial Variability and Uncertainty in Groundwater Flow Parameters: a Geostatistical Approach. *Water Resour. Res.* 15(2): 269-280.

Delhomme, J.P. (1978b). Kriging in Hydroscience. *Advances in Water Resources* 1(5): 251-266.

Dettinger, M.D. & Wilson, J.L. (1981). First Order Analysis of Uncertainty in Numerical Models of Groundwater Flow, 1, Mathematical Development. *Water Resour. Res.*, 17(1): 149-161.

Elfeki, A.M. (1991). *Mathematical Modelling of Seepage.* Unpublished Individual Study Report. International Institute for Hydraulic and Environmental Engineering, The Netherlands.

Elfeki, A.M. (1993). *Literature Review of Stochastic Simulation Models in Groundwater and Contaminant Transport.* Tech. Report # 355. Geotechnical Laboratory. Faculty of Civil Engineering. TU Delft. The Netherlands.

Elfeki, A.M. (1994a). *Practical Stochastic Methodology For Simulating Geological Formations from Soft Information Using Markov Chain Theory.* Tech. Report # 368, Geotechnical Laboratory. Faculty of Civil Engineering. TU Delft. The Netherlands.

Elfeki, A.M. (1994b). *Numerical Simulation of Groundwater Flow in Heterogeneous Formations Described by Soft Information: Model Development and Numerical Experiments.* Tech. Report # 369. Geotechnical Laboratory. Faculty of Civil Engineering. TU Delft. The Netherlands.

Elfeki, A.M. & Uffink, G.J.M. & Barends, F.J.B. (1995). Stochastic Simulation of Heterogeneous Geologcal Formations Using Soft Information, with An Application to Groundwater. In *Groundwater Quality: Remediation and Protection, QG'95.* Edited by Kovar, K. and Krasny, IAHS Publication No. 225.

Elfeki, A.M. & Uffink, G.J.M. & Barends, F.J.B. (1996). Solute Transport in Single and Multiple Scale Heterogeneous Formations: Numerical Experiments. Accepted for publications at *geoENV 96, First European Conference on Geostatistics for Environmental Applications,* Lisbon.

Elfeki, A.M. & Uffink, G.J.M. & Barends, F.J.B. (1996). Prediction Uncertainty of Contaminant Transport in Heterogeneous Formations: Numerical and Laboratory Experiments (In preparation).

Everitt, B.S. & Hand, D.J. (1981). *Finite Mixture Distributions.* Chapman and Hall. p. 143.

Farmer, C.L. (1988). The Generation of Stochastic Fields of Reservoir Parameters with Specified Geostatistical Distributions. Proceedings of The IMA Conference *'The Mathematics of Oil Production'.* Oxford University Press.

Fayers, F.J. & Hewett T.A. (1992). Review of Current Trends in Petroleum Reservoir Description and Assessing The Impact on Oil Recovery. *Computational Methods in Water Resources IX Conference. Vol. 2. Mathematical Modelling in Water Resources.* Computational Mechanics publications. Southhampton. Boston.

Fischer, H.B. & List, E.J. & Koh, C.Y. & Imberger, J. & Brooks, N.H. (1979). *Mixing in Inland and Coastal Waters.* Academic Press, New York.

Freeze, R.A. (1975). A Stochastic-Conceptual Analysis of One-dimensional Groundwater Flow in Nonuniform Homogeneous Media. *Water Resour. Res.* 11(5): 725-741.

Freeze, R.A. (1977). Probabilistic One-dimensional Consolidation. *J. of Geotechnical Engineering Division, ASCE.* Vol.(103), GT7, pp. 725-742.

Freeze, R.A. (1975). A Stochastic-Conceptual Analysis of One-dimensional Groundwater Flow in Nonuniform Homogeneous Media. *Water Resour. Res.* 11(5): 725-741.

Freeze, R.A. & Marsily G., de & Smith, L. & Massmann, J. (1987). Some Uncertainties about Uncertainty. In *Geostatistical, Sensitivity, and Uncertainty Methods for Ground-Water Flow and Radionuclide Transport Modeling.* Edited by Buxton, B.E. San Francisco, California, pp. 231-260.

Freyberg, D.L. (1986). A Natural Gradient Experiment on Solute Transport in a Sand Aquifer, 2, Spatial Moments and the Advection and Dispersion of Nonreactive tracers. *Water Resour. Res.* 22(13): 2031-2046.

Gelhar, L.W. (1977). Effects of Hydraulic Conductivity Variations on Groundwater Flow. In (Eds. Hjort, P., Johnsson, L. & Larsen, P.) *Proc. Int. Symp. Stochastic Hydraulics, 2nd Int. Assoc. Hydraulic Res.*, Lund, Sweden. Water Res. Pub. Fort Collins, Collins, Colorado, pp. 409-428.

Gelhar, L.W., Gutjahr, A.L. & Naff, R.L. (1979). Stochastic Analysis of Macrodispersion in a Stratified Aquifer. *Water Resour. Res.* 15(6): 1387-1397.

Gelhar, L.W. & Axness, C.L. (1983). Three-dimensional Stochastic Analysis of Macrodispersion in Aquifers. *Water Resour. Res.* 19(1): 161-180.

Gelhar, L.W. (1984). Stochastic Analysis of Flow in Heterogeneous Porous Media. In *Fundamental of Transport Phenomena in Porous Media*. Edited by J. Bear and M.Y. Corapcioglu, pp. 673-718, M. Nijhoff, Dordrecht, The Netherlands.

Gelhar, L.W. (1986). Stochastic Subsurface Hydrology: from Theory to Applications. *Water Resour. Res.* 22(9): 135S-145S.

Gelhar, L.W. & Welty, C. & Rehfeldt, K.R. (1992). A Critical Review of Data on Field-scale Dispersion in Aquifers. *Water Resour. Res.*, 28(7), pp. 1955-1974.

Gelhar, L.W. (1993). *Stochastic Subsurface Hydrology*. Prentice Hall, Englewood Cliffs.

Ghori, S.G., Heller, J.P., and Singh, A.K. (1992). An Efficient Method of Generating Random Fields. *J. Mathematical Geology*.

Glimm, J. & Lindquist, W.B. & Pereira, F. & Zhang, Q. (1993). A Theory of Macrodispersion for The Scale-up Problem. *Transp. Porous Media*. Vol. 13, pp. 97-122.

Gomez Hernandez, J.J. (1991). *A Stochastic Approach to The Simulation of Block Conductivity Values Conditioned Upon Data at a Smaller Scale*. PhD Thesis. Stanford University.

Gomez-Hernandez, J.J. & Gorelick, S.M. (1988) Influence of Spatial Variability of Aquifer and Recharge Properties in Determining Effective Parameter Values. In: Peck, A. et al. (eds.) *Consequences of Spatial Variability in Aquifer Properties and Data Limitations for Groundwater Modelling Practice*. IAHS publication 175, 217-272, IAHS Press, Oxfordshire.

Gottfried, B. (1984). *Elements of Stochastic Process Simulation*. Prentice Hall, Inc., Englewood Cliffs, New Jersey 07632.

Gutjahr, A.L. (1989). *Fast Fourier Transforms for Random Field Generation*. Project Report for Los Alamos grant to New Mexico Tech. Contract number 4 R58 - 2690R.

Gutjahr, A.L., Gelhar, L.W., Bakr, A.A. & McMillan, J.R. (1978). Stochastic Analysis of Spatial Variability in Subsurface Flows. Part II: Evaluation and Application. *Water Resour. Res.* 14(5): 953-960.

Gutjahr, A.L., & Gelhar, L.W. (1981). Stochastic Models of Subsurface Flow: Infinite Versus Finite Domain and Stationarity. *Water Resour. Res.* 17(2): 337-350.

Haldorsen, H.H., Brand, P.J., MacDonald, C.J. (1987). Review of The Stochastic Nature of Reservoir. In *Mathematics in Oil Production*. Eds. S. Edwards and P.R. King. Oxford Science Publications.

Haldorsen, H.H. (1983). *Reservoir Characterization Procedures for Numerical Simulation*. PhD. dissertation. University of Texas, Austin.

Haldorsen, H.H., Brand, P.J., Macdonald, C.J. (1987). Review of the Stochastic Nature of Reservoir. In *Mathematics in Oil Production*. Eds. S. Edwards and P.R. King. Oxford Science Publications.

Haldorsen, H.H. & Damsleth, E. (1990). Stochastic Modeling. *Society of Petroleum Engineers. SPE 20321*. Distinguished Author Series.

Hammersly, J.M. and Handscomb, D.C. (1964). *Monte-Carlo Methods*, Methuen, London.

Harbaugh, J.W. & Bonham-Carter (1970). *Computer Simulation in Geology*, J. Wiley, New York.

Harleman, D.R.F. & Melhorn, P.F. & Rumer, R.R. (1963). Dispersion permeability Correlation in Porous Media. *J. Hydraulic Div. ASCE*, 67-85.

Hemmerle, W.J. (1967). *Statistical Computations on a Digital Computer*. Blaisdell publishing company.

Hewett, T.A. (1986). Fractal Distributions of Reservoir Heterogeneity and Their Influence on Fluid Transport. *Paper SPE 15386*, presented at the Sixty-First Annual Technical Conference. Society of Petroleum Engineers. New Orleans.

Johnson, M.E. (1987). *Multivariate Statistical Simulation*. John Wiley & Sons, Inc. Journel, A.G. (1974). Geostatistics For Conditional Simulation of Ore Bodies. *Economic Geology*, 69(5): 673-687.

Journel, A.G. & Huijbregts, C. (1978). *Mining Geostatistics*. New York: Academic Press, 600p.

Journel, A. (1989). Fundamentals of Geostatistics in Five Lessons. *American Geophysical Union Press*. Washington D4. 40p.

Journel, A.G., Deutsh, C.V. & Desbarats, A.J. (1986). Power Averaging for Block Effective Permeability. *SPE 15128*, 56th. California regional meeting of SPE.

Jussel, P.F. & Stauffer, F. & Dracos, T. (1994a). Transport Modeling in Heterogeneous Aquifers: 1. Statistical Description and Numerical Generation of Gravel Deposits. *Water Resour. Res.* 30(6): 1803-1817.

Jussel, P.F. & Stauffer, F. & Dracos, T. (1994b). Transport Modeling in Heterogeneous Aquifers: 2. Three-dimensional Transport Model and Stochastic Numerical Tracer Experiments. *Water Resour. Res.* 30(6): 1819-1831.

Kay, S.M. (1988). *Modern Spectral Estimation*. Prentice Hall.

Kennedy, Jr.W.J., & Gentle, J.E. (1980). *Statistical Computing*. Marcel Dekker. New York.

Killy, R.W.D. & Moltyaner, G.L.(1988). Twin Lake Tracer Test: Methods and Permeabilities, *Water Resour. Res.* 24(10), pp. 1585-1613.

Kinzelbach, W. (1986). *Groundwater Modelling*. Developments in Water Science 25. Elsevier.

Kinzelbach, W. & Uffink, G.J.M. (1989). The Random Walk Method and Extensions in Groundwater Modelling. In Corapcioglu, M.Y. *Transport Process in Porous Media*. NATO ASI Series.

Koonin, S.E. and Meredith, D.C. (1990). *Computational Physics*. Addison-Wesley publishing company, Inc.

Kreft, A. & Zuber, A. (1978). On The Physical Meaning of The Dispersion Equation and Its Solution for Different Initial and Boundary Conditions. *Chem. Eng. Sci.*, 33: pp 1471-1480.

Krumbein, W.C. (1967). Fortran Computer Program for Markov Chain Experiments in Geology. *Computer Contribution 13*, Kansas Geologic Survey, Lawrence, Kansas.

Krumbein, W.C. (1970). Geological Models in Transition in Geostatistics. In *Geostatistics* Ed. by D.F. Merriam.

Lin, C. & Harbaugh, J.W. (1984). *Graphical Display of Two and Three Dimensional Markov Computer Models in Geology*. Van Nostrand Reinhold, New York.

Lippman, M.J. (1973). *Two-dimensional Stochastic Model of a Heterogeneous Geologic System*. PhD. Thesis. University of California, at Berkeley, California.

Lumb, P. (1966). The Variability of Natural Soils. *Canadian Geotechnical Journal*, vol.(3), no. 2, pp. 74-97.

Lumley, J.L., and Panofsky, H.A. (1964). *The Structure of Atmospheric Turbulence*. J. Wiley, New York.

Mantoglou, A. & Wilson, J.L. (1982). The Turning Bands Method for Simulation of Random Fields Using Line Generation by Spectral Method. *Water Resour. Res.* 18(5): 1379-1394.

Mantoglou, A. (1987). Digital Simulations of Multivariate Two- and Three-dimensional Stochastic Processes with a Spectral Turning Bands Method. *Math. Geology.* 19(2): 129-149.

Marsily, G. de (1986). *Quantitative Hydrogeology: Groundwater Hydrology for Engineers.* Academic Press, Orlando, FL, pp. 28-266.

Matheron, G. (1971). *Theory of Regionalized Variables and Its Applications.* 211pp., Ecole des Mines. Fontainebleau. France.

Matheron, G. & Marsily, G. de (1980). Is Transportation in Porous Media Always Diffusive? A Counter Example. *Water Resour. Res.* 16(5): 901-917.

Matalas, N.C. (1967). Mathematical Assessment of Synthetic Hydrology. *Water Resour. Res.* 3(4): 937-945.

Mejia, J.M. & Rodriguez-Iturbe I. (1974). On The Synthesis of Random Field Sampling from The Spectrum: An Application to The Generation of Hydrologic Spatial Processes. *Water Resour. Res.* 10(4): 705-712.

Merriam, D.F. (1976). *Random Processes in Geology.* Springer-Verlag, New York.

Mercado, A. (1967). The Spreading Pattern of Injected Water in Permeability-Stratified Aquifer. *Symposium of Haifa, Artificial Recharge and Management of Aquifers.* IASH Publ. 72, 23-36.

Mood, A.M. and Graybill, F.A. (1963). *Introduction to The Theory of Statistics.* McGraw-Hill Book Co., Inc., New York.

Neuman, S.P. (1980). A Statistical Characterisation of Aquifer Heterogeneities: An Overview. In (Ed. Narasimhan, T.N.) *Recent Trends in Hydrogeology. Spec. Pap. Geol. Soc. Amer.* 189: 81-102. Boulder, Colorado.

Neuman, S.P. (1984). Role of Geostatistics in Subsurface Hydrology. In (Eds. Verly, G., David, M., Journel, A.G. & Marechal, A.) *Geo-statistics for Nature Resources Characterization.* Proc. NATO-ASI, Part 1, pp. 787-816. Reidel, Dordrecht, The Netherlands.

Neuman, S.P. & Winter, C.L. & Newman, C.M. (1987). Stochastic Theory of Field-scale Fickian Dispersion in Anisotropic Porous Media. *Water Resour. Res.* 23(3): 453-466.

Parker, J.C. & van Genuchten, M.TH. (1984). Flux-averaged and Volume-averaged Concentrations in Continuum Approaches to Solute Transport. *Water Resour. Res.,* 20, pp. 866-872.

Peck, A., Gorelick, S., Marsily, G. de, Foster, S. & Kovalevsky, V. (1988). *Consequences of Spatial Variability in Aquifer Properties and Data Limitations for Groundwater Modelling Practice.* IAHS publication 175, IAHS Press, Oxfordshire.

Ripley, B.D. (1981). *Spatial Statistics,* J. Wiley, & Sons, Inc.

Ripley, B.D. (1987). *Stochastic Simulation,* J. Wiley & Sons, Inc.

Rubinstein, R.Y. (1981). *Simulation and the Monte-Carlo Method.* J. Wiley, New York.

Sahimi, M. (1993). Fractal and Superdiffusion Transport and Hydrodynamic Dispersion in Heterogeneous Porous Media, *Transp. Porous Media,* 13, pp.3-40.

Sahimi, M. (1995). *Flow and Transport in Porous Media and Fractured Rocks: From Classical Methods to Modern Approaches.* VCH Verlagsgesellschaft. mbH, D-69451 Weinbeim.

Serrano, S.E. & Unny, T.E. (1987). Semigroup Solutions of The Unsteady Groundwater Flow Equation with Stochastic Parameters. *Stoch. Hydrol. Hydraul.,* 1, pp. 281-296.

Serrano, S.E. (1988). General Solution of Random Advection-dispersive Transport Equation in Porous Media, 1. Stochasticity in The Sources and in The Boundaries. *Stoch. Hydrol. Hydraul.,* 2, pp. 79-98.

Serrano, S.E. (1988). General Solution of Random Advection-dispersive Transport Equation in Porous Media, 2. Stochasticity in Parameters. *Stoch. Hydrol. Hydraul.,* 2, pp. 99-112.

Schwartz, F.W. (1977). Macroscopic Dispersion in Porous Media: The Controlling Factors. *Water Resour. Res.*, 13(4), pp. 743-752.

Shinozuka, M. & Jan, C.M. (1972). Digital Simulation of Random Processes and Its Applications. *J. of Sound and Vibration*, 25(1): 111-128.

Shinozuka, M. (1971). Simulation of Multivariate and Multidimensional Random Processes. *J. Acoust. Soc. Am.*, vol.(49), no. 1 (part 2), pp. 357-367.

Smith, L. & Freeze, R.A. (1979a). Stochastic Analysis of Steady State Groundwater Flow in a Bounded Domain. 1. One-dimensional Simulations. *Water Resour. Res.* 15(3): 521-528.

Smith, L. & Freeze, R.A. (1979b). Stochastic Analysis of Steady State Groundwater Flow in a Bounded Domain. 2. Two-dimensional Simulations. *Water Resour. Res.* 15(6): 1543-1559.

Smith, L. & Schwartz, F.W. (1980). Mass Transport. 1. Analysis of Macrodispersion. *Water Resour. Res.* 16(2): 303-313.

Smith, L. & Schwartz, F.W. (1981a). Mass Transport. 2. Analysis of Uncertainty in Prediction. *Water Resour. Res.* 17(2): 351-369.

Strikwerda, J.C. (1989). *Finite Difference Schemes and Partial Differential Equations.* Wadsworth & Brooks / Cole Advanced Books & Software.

Sudicky, E.A. (1986). A Natural Gradient Experiment on Solute Transport in a Sand Aquifer: Spatial Variability of Hydraulic Conductivity and Its Role in Dispersion Process. *Water Resour. Res.*, 22(13): 2069-2082.

Tang, D.H. & Pinder, G.F. (1977). Simulation of Groundwater Flow and Mass Transport Under Uncertainty. *Advances in Water Resources* 1(1): 25-30.

Titterington, D.M. & Smith, A.F. & Makov, V.E. (1985). *Statistical Analysis of Finite Mixture Distribution.* John Wiley and Sons. 243 p.

Tompson, A.F.B. & Ababou, R. & Gelhar, L.W. (1989) Implementation of The Three-dimensional Turning Bands Random Field Generator. *Water Resour. Res.* 25(10): 2227-2243.

Tompson, A.F.B. & Gelhar, L.W. (1990). Numerical Simulation of Solute Transport in Three-dimensional, Randomly Heterogeneous Porous Media. *Water Resour. Res.* 26(10): 2541-2562.

Tong, Y.L. (1990). *The Multivariate Normal Distribution.* Springer-Verlag. New York Inc.

Townley, L.R. (1983). *Numerical Models of Groundwater Flow: Prediction and Parameter Estimation in The Presence of Uncertainty.* PhD Thesis. Dept. of Civil Engineering. M.I.T., Cambridge, Massachusetts.

Townley, L.R. (1984). Second Order Effects of Uncertain Transmissivities on Prediction of Piezometric Heads. In Proc. 5th Intern. Conf. on *Finite Elements in Water Resources*, Burlington, Vermont: Springer-Verlag.

Townley, L.R. & Wilson, J.L. (1985). Computationally Efficient Algorithms for Parameter Estimation and Uncertainty Propagation in Numerical Models of Groundwater Flow. *Water Resour. Res.* 21(12): 1851-1860.

Townley, L.R. (1988). Modelling Flow in Heterogeneous Aquifers: Identification of The Important Scales of Variability. In: *Developments in Water Science 35.* Proc. International Conference, MIT, USA, vol.(1) pp. 197-202. Elsevier.

Uffink, G.J.M. (1990). *Analysis of Dispersion by The Random Walk Method.* Ph.D. Thesis. Delft University of Technology. The Netherlands.

Vanmarcke, E. (1977). Probability Modelling of Soil Profiles: Proc. *ASCE. Jour. of Geotechnical Eng. Div.* Vol.(103) No. GT11, pp. 1227-1247.

Vanmarcke, E. (1983). *Random Fields: Analysis and Synthesis.* Cambridge, Mass: MIT Press.

Vomocil, J.A. (1965). 'Porosity', in Methods of Soil Analysis. Pt. I., C.A. Black, ed., *American Society of Agronomy*, Madison, WI, 229 p.

Warren, J.E., and Price, H.S. (1961). Flow in Heterogeneous Porous Media. *Soc. Pet. Eng. J.*, vol.(1), pp. 153-169.

Warren, J.E., and Price, H.S. (1964). Macroscopic Dispersion: Transactions AIME, *Soc. Pet. Eng. J.*, vol.(231), pp. 215-230.

Weber, K.J. (1986). How Heterogeneity Affects Oil Recovery. *Reservoir Characterization*. Academic Press, NY. pp. 487-544.

Westlake, J.R. (1968). *A Handbook of Numerical Matrix Inversion and Solution of Linear Equations*. Control Data Corporation, USA.

Wheatcraft, S.W. & Tyler, S.W. (1988). An Explanation of Scale-dependent Dispersivity in Heterogeneous Aquifers Using Concepts of Fractal Geometry. *Water Resour. Res.* 24 (4), pp. 566-578.

Whittle, P. (1954). On Stationary Processes in The Plane. *Biometrika,* 41, pp. 434-449.

Woodbury, A.D. & Sudicky, E.A. (1991). The Geostatistical Characteristics of The Borden Aquifer. *Water Resour. Res.*, 27(4), pp. 533-546.

Yevjevich, V. (1972). *Stochastic Processes in Hydrology*. Water Resources Publications. Fort Collins, Colorado, USA.

Zeitoun, D.G. & Braester, C. (1991). A Neumann Expansion Approach to Flow Through Heterogeneous Formations. *Stoch. Hydrol. Hydraul.*, 5, pp. 207-226.

9 789054 106654